THE DESCENT OF HUMAN SEX RATIO AT BIRTH

METHODOS SERIES

VOLUME 4

This Book Series is devoted to examining and solving the major methodological problems social sciences are facing. Take for example the gap between empirical and theoretical research, the explanatory power of models, the relevance of multilevel analysis, the weakness of cumulative knowledge, the role of ordinary knowledge in the research process, or the place which should be reserved to "time, change and history" when explaining social facts. These problems are well known and yet they are seldom treated in depth in scientific literature because of their general nature.

So that these problems may be examined and solutions found, the series prompts and fosters the setting-up of international multidisciplinary research teams, and it is work by these teams that appears in the Book Series. The series can also host books produced by a single author which follow the same objectives. Proposals for manuscripts and plans for collective books will be carefully examined.

The epistemological scope of these methodological problems is obvious and resorting to Philosophy of Science becomes a necessity. The main objective of the Series remains however the methodological solutions that can be applied to the problems in hand. Therefore the books of the Series are closely connected to the research practices.

The titles published in this series are listed at the end of this volume.

The Descent of Human Sex Ratio at Birth

A Dialogue between Mathematics, Biology and Sociology

Éric Brian
Ecole des Hautes Etudes en Sciences Sociales, Paris, France

and

Marie Jaisson
University Francois-Rabelais, Tours, France

Springer

A C.I.P. Catalogue record for this book is available from the Library of Congress.

ISBN 978-90-481-7511-6
ISBN 978-1-4020-6036-6 (e-book)

Published by Springer,
P.O. Box 17, 3300 AA Dordrecht, The Netherlands.

www.springeronline.com

Printed on acid-free paper

CONTENTS

FOREWORD

This book has been written for the critical attention of researchers who rarely work together: mathematicians, biologists, historians, philosophers, social scientists and historians of the sciences. Its object – how close the numerical ratio of the two sexes at birth may come to some regularity – is in fact almost three centuries old, which is more than the collective memory of any of these disciplines or the individual consciousness of the specialist scholar can generally envisage. But whether we like it or not, if the conclusion were to be drawn tomorrow that a fairly stable proportion could be measured (perhaps the proportion of the two sexes, perhaps another), then it would be one of the epistemological registers examined in our enquiry that would come into play. Sometimes figures seem to have too much to say for themselves; they happen to give off the musty smell of theology, to lead readers towards the evasion of phenomena that are actually relevant, to draw them into the twists and turns of uncontrolled philosophies of history and philosophies of sciences... It then falls to those who have made science their profession to exercise a threefold control over figures: control through the technical coherence of formal methods of calculation, control by matching the conceptual analytical devices to the object as studied, and control through the relevance of the intellectual genealogies that these methods and these concepts involve. Clearly, since the book will be read from the point of view of more than one discipline, these controls will be exercised in different ways. This does not matter, as long as, when all is said and done, this threefold control remains manifestly cogent every time.

The plan for this book arose when several contextual factors came together. At the end of the academic year 2002–2003, following discussions held at the Institute for Advanced Studies in Berlin (the *Wissenschafts-kolleg*), we were left wanting to go further into a dialogue between the social sciences and the biological sciences. We felt it was indispensable to locate objects likely to stimulate such encounters (like the analysis of the proportion of the two sexes) and then move on to take a serious look at previous states of tension between the disciplines around these objects – tensions that might be both very old and very much alive. All these things would be difficult to put across in a symposium, much easier to introduce through a book. A book could also draw on re-examinations of research on the history of social sciences in the 18th and 20th centuries, in which we had recently taken part. Finally, a book would be able to respond to a critical set of circumstances peculiar to the social sciences by suggesting the possibility of a new conception of the long historicity of the sciences, and consequently the opportunity for re-examination at the borders of other disciplines, most particularly where mathematical thinking or biological facts are involved. At this point, our project received the attention and encouragement of Daniel Courgeau and Robert Franck: without them, the methodological purpose of this study would probably not have found the more favourable conditions that have allowed it to take the concrete form of a book.

The *Descent of human sex ratio at birth* follows and extends another book that we prepared together, in which several other authors also collaborated: the critical edition of a very little-known text published in 1936 by Maurice Halbwachs and Alfred Sauvy, *Le Point de vue du nombre* (published by INED, Paris, 2005). In parallel, we have also written a sociological work that aims to define the stochastic nature of social phenomena through a case study: *Le Sexisme de la première heure* (Raisons d'agir, Paris, 2007). Between them, these three works are interrelated in the same way as – to evoke a famous precedent on which we shall comment later – the Introduction to the *Théorie analytique des probabilités* (1812) and the *Essai philosophique sur les probabilités* (1814). Some elements of the first are taken up and extended in the later works. However, each has a different objective: *Le Point de vue du nombre* (2005) gives an account of a collective scholarly enquiry, addressed to historians of the social sciences; with *Le Sexisme de la première heure* (2007), arguments are presented to social scientists in favour of identifying a new sociological object and demonstrating the relevance of a new approach; and here, methodical thinking that crosses several disciplines – different forms of mathematics, biology, the social sciences – is submitted to the judgment of specialists and of epistemologists. The three volumes together come from the same work in progress, a project whose aim is the reflexive re-examination of the social sciences.

After our stay in Berlin, several stages in our research were discussed within an institutional triangle in Paris, marked out by the beacons of the *Institut national d'études démographiques* (the National Institute for

Demographic Studies - INED), the *École des hautes études en sciences sociales* (the School for Advanced Studies in Social Sciences - EHESS) and the completely new *École d'économie de Paris* (the Paris School of Economics - PSE). Discussions with our colleagues at the *Centre Maurice-Halbwachs* (CNRS-EHESS-ENS-UCBN-EEP) and the *History and Populations* research unit (INED) were always particularly fruitful. In addition, the publication of this book, and notably its translation, would not have been possible without INED funding and without the support of its Director, François Héran, himself active in these scientific discussion networks. We are also very grateful to Karen George for the quality of her work in translating and fine-tuning the English manuscript, and for the relevant exchanges that we had with her in the course of this work. Finally, Catriona Dutreuilh, Translation Coordinator at INED, and Evelien Bakker, Associate Publishing Editor for Social Sciences at Springer, followed the preparation of this publication considerately and efficiently. We hope that everyone who has played a direct or indirect part in the preparation of this book will find that its publication brings echoes of the intense moments we have shared – which, for us, were indispensable.

Paris, October 2006

NOTE ON TRANSLATION

In working on this book, we have found ourselves following Ariadne's thread to a series of clues that bind the calculation of the proportion of boys and girls at birth to a body of works that have appeared in various places, in various languages – English, German, French, Italian, Latin – and at various dates from the 18th to the 21st centuries... In doing so, we have had to assess the important part played in the formation of contemporary sciences by the – sometimes long-standing – circulation of published works, of words, of methods and of indices, of the intellectual transfers that have accompanied this circulation, and most particularly of the variations, ambiguities and shifts inherent in them – all 'failings' stigmatized by academic standards from a regulatory point of view, but where a virtue can be made of necessity only as long as they pass unnoticed by the critics

In these conditions, it is important to make clear to the potential reader of this work – we imagine a student or a scientist, conversant with the English language in its early-21st-century international written form – that it was first written in French by authors who regard this language as their critical working tool: that is, they situate themselves within intellectual traditions consolidated in France during the 20th century (a claim of social science maintained throughout that century and marked by the activities of three journals – Emile Durkheim's *L'Année sociologique*, Henri Berr's *La Revue de synthèse* and Lucien Febvre's and Marc Bloch's *Les Annales*; by an attention to epistemology, in the wake of Gaston Bachelard's works; and by participation in current renewals of the social sciences). The authors' next step was to establish a second version of the manuscript, close to the first but conceived specifically for an international readership. The third stage was for this new version to be translated into English – targeted at our student or scientist, whether native English speaker or well-read non-native specialist – while recognizing the impossibility and inappropriateness of rendering scholarly French in the common denominator of "English as a Foreign Language". It was decided to adopt the "Oxford" spelling conventions preferred in most British English academic publications, conventions which are most comfortable for an international readership. Detailed discussions between the authors and the translator then helped to perfect the resulting manuscript. Finally, the translator established the definitive text, to which a few proofreading corrections have subsequently been made.

Let us clarify the perspective adopted by the authors and the translator. Since each chapter deals with a particular place and moment, the context and the vocabulary of that era and that setting have governed the choice of expressions used. It is important to combat the effects of anachronistic reformulations, which very often arise whenever authors from days gone by are mentioned. Without entering into a case-by-case discussion – which would overload this Note and, in passing, open any number of Pandora's boxes for the historian of sciences and the rigorous scientist to delve into – let us simply say that the vocabulary of objectivization, of calculation, of estimation, of probabilities and of statistics has here been chosen to be as close as possible to that of the authors commented on and has as far as possible been verified in the historical archive of the Oxford English Dictionary. With the best-known of these authors who did not write in English, the translation of their works (where this exists by way of a standard edition) often appeared at a much later date than the initial publication, or else became established in an intellectual context somewhat foreign to the one in which the work was produced. This is most particularly true for Condorcet, Laplace, Comte and Durkheim. Therefore we have adopted the policy of relying on available standard translations, while correcting them where they have departed too much from our requirement to restore the original. (The variations that we offer our reader appear in italics in the extracts concerned, and are duly acknowledged.)

Beyond these general indications, we should add that we found two groups of expressions hard to untangle. The first group is marked out by the English expressions *human race, human species, humanity, human beings, mankind* and *humankind* and their French counterparts *l'espèce humaine, l'humanité, les êtres humains, l'Homme, le genre humain*. Each of these terms would require a historical semantic study that has no place here (as this book appears, work is already going on in this field in several parts of Europe). We have preferred the use of *humankind* in order to express what is meant in French – at the time of the Enlightenment, in Condorcet for example, or even today – by the words *l'espèce humaine* or *l'humanité*: that is, *human beings* as a whole considered from a normative point of view and without distinction or connotation of race, condition or gender.

To many English readers, the word *humankind* may appear to be a 20th-century artefact, contrived to meet a particular purpose: this is far from the case – it has an honourable pedigree, attested in the 17th century from the Restoration poets and in the 18th from Alexander Pope. Throughout our text, therefore, *humankind* is used by design, and any other variations that appear have been chosen deliberately, taking into account any connotations they may carry.

A second group of expressions relates to the system of qualifying *male/female* in English in order to distinguish the sexes or the genders. In English, this vocabulary is common to the description of human beings and the description of other animals. In French, *homme/femme* or else *masculin/féminin* is used when referring to human beings, with *mâle/femelle* generally reserved for other animals. The use of the latter about human beings in French would indicate the deliberate choice of a degrading vocabulary. As we shall see, for the issues that we are opening up here, we must take seriously the fact that some authors view the analogy between *humankind* and the other *animal species* as obvious, others as unfounded and others again as problematic. Therefore it is important to proscribe the unthinking use of *male/female* when exploring and reconstructing the body of work that forms the object of our study. The authors would like to add that – still themselves deeply shocked at the use of the terms *male* or *female* in relation to human beings – they are pleased to take comfort in knowing that this proscription has been followed in the translation of their work.

Éric Brian, Marie Jaisson, Karen George.
December 2006.

INTRODUCTION

It would be unwise to reopen the issue of analysis of the proportion of the sexes at birth without taking to heart Corrado Gini's warning, given a century ago, laying bare the futility of such a quest. From the very first pages of his thesis (1908), the statistician Gini (1884–1965) invoked his distant predecessors: the French doctor who worked in Leiden, Charles Drelincourt (1633–1697), and the Göttingen naturalist, Johann Friedrich Blumenbach (1752–1840):

> "Even before the 19th century, Drelincourt had already listed 262 'unfounded' hypotheses on the nature and cause of sex, while Blumenbach had observed acerbically that there was nothing to prove that Drelincourt's own hypothesis was not No 263. Since then, the number of hypotheses has more than doubled – including, of course, Blumenbach's own theory of *Bildungstrieb*. And this number continues to grow day by day." (Gini, 1908, pp. 8–9)

We shall be discussing the calculation of the proportion of the sexes at birth in human beings, but not "the cause of the sexes", which has been debated in vain since Antiquity. Consequently, our work will cover almost three centuries of studies, as it is only since the early 18th century that scholars have taken this proportion as an object of research. From that time forward, it has often been regarded as almost constant. Nowadays, following Ronald A. Fisher (1890–1962), the view is taken that the trend of the sex ratio is, in principle, to adjust towards a balance of the two sexes. However, the persistent, exactly equal difference in this proportion at birth, and even its relatively small variations have provided food for thought for theologians, mathematicians, social scientists and biologists ever since the first calculations were made, even up to the present day. At the end of the 19th century, the economist and statistician Francis Y. Edgeworth (1845–1926) noted that the phenomenon of the regularity of the ratio between the two sexes at birth lent itself to academic approaches and constructions relevant to various disciplines, but irreducible from one discipline to another.

"The familiar observation that the areas of art and science do not coincide is nowhere more strikingly exemplified than in [this] field of inquiry [...]. The fact that some five per cent more boys are born than girls is probably a mere *curiosum* in the eyes of the practitioner; yet it has a theoretical value for the biologist, especially when compared with similar observations for the inferior animals, and even plants [...]. Moreover, even if the fact were entirely isolated and remote from physiological inquiries, its investigation would still possess a scientific interest, as affording a particularly perfect study in statistical method." (Edgeworth, 1892, pp. 337–338)

Yet recently, calculations for various regions of the world have highlighted large disparities. In rich countries, about 51.2% of births are of boys and 48.8% of girls, which is about 105 boys per 100 girls (as Edgeworth indicated), while in China the equivalent figures are in the order of 55%, 45% and 122 per 100... In these very recent conditions, work on human sex ratio at birth has seen a real resurgence, now combining new contributions from demography, economics and biology. Therefore it is important nowadays to question what may appear to be an invariable feature of the study of sex ratio: for three centuries, it has always brought into opposition various established disciplines of the time. In our eyes, a true renewal of research on this terrain calls for a critical evaluation of what therefore seems to be a *de facto* given, specific to the conditions of knowledge of the phenomena captured by the ratio. It must be accompanied by an epistemological and methodological diagnosis of whether this state of affairs is necessary, and – why not? – for proposals that might offer new ways of reconstructing the phenomenon.

How could we construct a critical perspective that would allow us to include three entire centuries of "conflicts of the faculties" about the degree of regularity of the proportion of the sexes at birth? The history of the calculation itself would not teach us much: it is by and large a matter of a simple ratio of one number to another... Throughout these three centuries, it has always been known how to establish this calculation. Yet it has constantly been understood in different ways: as the comparative state of enumerations, as the arithmetical ratio, as the comparison of the chances of one or the other of the two sexes being born, as relative frequency, as the measure of a greater facility of the birth of one of the two sexes, as the estimate of a probability of one sex at birth, as the comparative assessment of the probabilities of survival of the two sexes... all these conceptual developments are compatible with the same ratio. The history of the calculation process cannot be divorced from the history of these developments, and we find ourselves faced with transformations in shifting configurations of academic specialisms in various eras. The historiographies of mathematical, biological or sociological ideas, for exactly symmetrical reasons, would not offer any help if we were to consider them individually.

So the policy we shall adopt is to study various moments in the history of the uses of a formal method – the calculation of the proportion of the sexes at birth – while also highlighting the tensions between the academic specialisms involved in its interpretation and reconstructing how, from one period to another, the products of these old tensions have been integrated into the collective memories of later specialisms. In doing so, for each period that we consider, we shall seek to highlight several elements for analysis. The first of these will be the objects to which scholars devoted themselves in former times: for example, humanity understood as a whole, the human species viewed by implicit or explicit comparison with animal or plant species, populations at the national or other scales, social groups. Secondly, we shall seek to account for the dependence of scholars on empirical material: registers, compilation operations, calculation methods, the handling of uncertainties and errors. A third category of issues will also guide our study, that of the institutional frames in which work on the proportion of the sexes at birth has been situated: the backdrop of academic institutions (both scholarly societies and universities) and the disciplinary claims specific to the intense quest from which the early works came. Finally, we shall seek to retrace as precisely as we can the conditions of transmission and the means of communication between the different works on which we are commenting. Taking stock of all this will allow us to give an outline of the history of the social division of labour of the calculation of the sex ratio that will illuminate the very specific conditions of each given time – conditions in which numerical abstraction was able to meet particular theoretical concepts, where what mattered at different times was, for example, the arguments of theologians or even naturalists, or a deterministic philosophy of the sciences, or a conception of probability.

This kind of outline cannot take the place of a general history – whether of statistics, of the social sciences, or even of relationships between areas such as the calculus of probabilities or biology! Indeed, it is completely impossible in the current state of historiography to deal in a rigorous manner with all the criteria that we have just listed consistently across three centuries, even though our topic is confined to the calculation of human sex ratio at birth. Each of our chapters will serve as a case study. The way we link them together and compare them will aim to provide the reader with as relevant a perspective as possible, enriched with historical elements that were already known or have been established through our research. Happily, a *de facto* given makes following this programme easier than it might seem at first glance. Over the three centuries, systematic sources and genuinely new work have been rare. Therefore, we have been able to organize this outline by determining to follow the fortunes of these rarities. That is why, throughout our study and our writing, we have paid careful attention to the concrete phenomena – whether legitimate or not – through which these resources have been transmitted and through which the forms taken by collective memory have been shaped in the groups of scholars concerned.

The first stage of our work is an assessment of how the scholarly world of the 18th century was able to conceive of regularity or variations in the proportion of the sexes at birth. Physico-theologians – such as John Arbuthnot (1667–1735) or Johann Peter Süssmilch (1707–1767) – and mathematicians involved in founding the analytical calculus of probabilities – Condorcet (1743–1794) or Laplace (1749–1827) – were the main players in the specialized discussions that took place within the principal learned societies of London, Berlin and Paris (Chapter 1). Secondly, in the early 19th century, there were agronomists – Charles Gilbert de Morel-Vindé (1759–1841) or Charles Girou de Buzareingues (1773–1856) –, physiologists – Johann Daniel Hofacker (1788–1828) –, organizers of statistical observations – Joseph Fourier (1768–1830) or Adolphe Quetelet (1796–1874) –, mathematicians – Fourier again, or Denis Poisson (1781–1840) – who all came into conflict around the relationships between birth figures in Paris, London, Tübingen and Brussels (Chapter 2). Then, in the mid-19th century – with Quetelet – the issue of compiling lists of birth registrations came to override the earlier debates. Different forms of production of statistics proliferated. The base provided by calculations was sealed away for a long time (Chapter 3). But in parallel, first in England, then at the turn of the 20th century in Germany and in Italy, a completely different conception of the variability of numerical indices and the conclusions that they give was consolidated in the biological literature – from Charles Darwin (1809–1882) to Ronald A. Fisher, through Francis Galton (1822–1911), Carl Düsing (born in 1859) and Corrado Gini (Chapter 4). Another route again, contemporary with the preceding one, leads to discussion of the regularity of the ratio of the sexes at birth through the sociological qualification of the empirical consistency of the social fact. This was specific to the French conception of sociology – notably of Auguste Comte (1798–1857), Émile Durkheim (1858–1917) and Maurice Halbwachs (1877–1945) (Chapter 5). But it is not enough just to give this historical deconstruction of the sex ratio as a kind of long history of the production and circulation of resources and traces of intellectual work. We end our various journeys, therefore, in one place, where the elements we have gleaned allow us to reconstruct some of the phenomena of which the sex ratio is an indicator (Chapter 6).

We have appended three texts to this series of case studies. The first is made up of extracts from a manuscript of Condorcet's written in 1793 or 1794, where the philosopher and mathematician discusses the effects of an intervention by humankind on the chances of one sex rather than the other being born (Appendix A). To our knowledge, this represents the first time that the trend of the sex ratio at birth to adjust towards balance between the sexes was expressed as a principle. Nowadays, a principle of this kind is most readily associated with the name of Fisher (1930), while some recent commentators have also mentioned the first edition of Darwin's *The Descent of Man* (1871) and the thesis of the physiologist Düsing (1883 and 1884) in relation to this topic. The second document offered as an appendix is well known, consisting of precisely those passages that Darwin devoted to the human sex ratio at birth in the first two editions of his book, published in 1871 and 1874. The fact that the great naturalist quickly revised his thinking is well known. This has been commented on several times, so it is useful to have the text itself and its variants to hand (Appendix B). The third document is as little known as the first. It consists of extensive extracts from an article published by Maurice Halbwachs in the *Journal de la société de statistique de Paris* in 1933, where he proposed to explain variations in the proportion of the sexes at birth following a sociological analysis (Appendix C). This endeavour has remained forgotten since the Second World War. Finally, a fourth appendix supplies points of reference for the indices of sex ratio and for their links with the calculus of probabilities (Appendix D).

Our conclusion reviews these contributions from the long perspective of three centuries, a perspective that gives a better view of the importance of the phenomena of the contextualized circulation of knowledge between constituted disciplines, between varied linguistic spaces and from one scientific moment to another. Today, these journeys lead us towards questions that can reasonably be addressed to each of the disciplines concerned: mathematics, biology and sociology.

Chapter 1

PHYSICO-THEOLOGY AND MATHEMATICS (1710–1794)

1. THE CENTURY OF PHYSICO-THEOLOGY

Our theme can be tackled initially through the history of ideas, although we shall have to move quickly away from this approach. In fact, its demographic treatment in 2001 was in keeping with the approach taken by Johann Peter Süssmilch when he first attempted to tackle the issue systematically, no less than 260 years earlier. A Lutheran pastor from Berlin, he wrote: "for every 1,000 baby girls born, there are 1,050 boys". In his eyes, this proportion governed the order of human reproduction; this, as we can see, makes the figure of 105 boys to 100 girls "one of the rare demographic parameters that are almost constant" (Caselli & Vallin, 2001, p. 57). Even the choice of index – the number of boys to every one hundred female births – is puzzling, although it is very widespread in the statistical, demographic and biological literature (Caselli, Vallin, & Wunsch, 2001–2004; Daguet, 2002; Hardy, 2002; Majerus, 2003). It is one of the rare proportions that is expressed in the language of the specialist – nowadays, as in the 18th century – as if we were talking about placing a bet, comparing 105 chances of giving birth to a boy to 100 chances of giving birth to a girl (Cournot, 1843, pp. 21–24). The terms of reference used over the last two centuries could just as well have been to the proportions of the two sexes in the total number of births (in other words, their frequencies) – or perhaps even to the calculus of probabilities (Laplace, 1778 [1781]).

To make comparisons easier, we prefer not to think in terms of 105 boys for 100 girls, but 105 boys out of 205 births – which is 51.2% – or its complement of 100 girls out of the same total, which is 48.8%. The collective attachment to a very old figure and an old-fashioned means of

E. Brian and M. Jaisson, The Descent of Human Sex Ratio at Birth, 1–25.

expression (one which is ill-founded on the mathematical level)[1] seems strange, at a time when historians of demography are devoting themselves to characterizing the 18th century – with its absence of modern register office procedures and its ignorance of 19th- and 20th-century administrative statistics – as the prehistory of the discipline (Dupâquier & Dupâquier, 1985). For either we must credit Süssmilch with inventing a quasi-constant for calculating populations and say that demography did not need the last two centuries of registration to calculate them, or else we must accept that Süssmilch and Vallin live in non-comparable – if not non-commensurable – worlds, and any similarity between their figures is meaningless. The dual absurdity of these alternatives should be enough to convince us of the need to look more closely at the Age of Enlightenment.

Although the causes that could help to determine the sex of a newborn infant have been the object of scholarly comment since Antiquity, it was in the 18th century that the regularity of an excess of male births drew the attention of naturalists and scholars. The positions taken on the question between 1740 and 1790 may be summarized by highlighting four distinct views, defended respectively by Johann Peter Süssmilch (1707–1767), a theologian and scholar active in the Berlin Academy of Sciences (Brig, 2001; Rohrbasser, 2001), Jean Le Rond D'Alembert (1717–1783), Pierre Simon de Laplace (1749–1827) and Marie Jean Antoine Nicolas de Caritat, Marquis de Condorcet (1743–1794), who were all mathematicians at the Paris Academy of Sciences, where the last two followed in the academic footsteps of the first – though not without some rivalry (Brian, 1994a).

In the 19th century, the historiography of the sciences and historical discourse on the social sciences consolidated their initial common framework around the view of a "general history of the sciences" – to the formation of which D'Alembert and Condorcet themselves directly contributed, and on which Auguste Comte (1798–1857) set the seal – and so for a long time they concealed a view, dominant among European

[1] If we let M be the number of boys, F the number of girls and $N = M + F$ their total, and if we consider that the probability of the sex of a child to be born is analogous to the toss of coin but with 51.2% probability on the side of the male sex and 48.8% on the side of the female sex, then – according to one result of the calculus of probabilities, to which we will return – the proportion M/N obeys a distribution well enough known for us to be able to write that its dispersal is measured by $1/2\sqrt{N}$. The relationship M/F is less simple to capture through the calculus of probabilities (since the numerator and the denominator are both uncertain). However we can estimate the order of magnitude of its dispersal. It is four times greater than the preceding measure (See Appendix D).

academic élites, notably in the Protestant countries, England, Prussia and Sweden, which may be covered by the term "physico-theology" (Derham, 1713). The Newtonian Anglican priest, astronomer and naturalist, William Derham (1657–1735), the doctors and mathematicians Bernard Nieuwentyt (1654–1718) – a Dutch Cartesian and anti-Spinozist – and John Arbuthnot (1667–1735) – an independently-minded Scot with a passion for natural philosophy – or, once again, the Leibnizian Lutheran pastor, Johann Peter Süssmilch, were among the most important names, with the last two examining more particularly the proportion of the sexes at birth. This tendency devoted itself on principle to the study of natural phenomena according to the standards of empirical scholarly work in the 18th century. This *physics* was thus admissible in the majority of the most well-known learned societies. Such research was directed at the manifestations of order willed by divine power: it was therefore also a *theology*. These authors intended to take the empirical route towards resolving the discussions of scholars and philosophers that were so much haunted by the supreme powers conceived of by Descartes (1596–1650), Spinoza (1632–1677), Newton (1643–1727) and Leibniz (1646–1716) (Rohrbasser, 2001).

Thus, for example, trained in the Lutheranism of the University of Halle and in the Leibnizianism of Christian Wolff (1679–1754), Süssmilch considered that divine providence was continually at work in the formation of every being, thus manifesting its continual calculations and ensuring the order appropriate for it (Rohrbasser in Süssmilch, 1998). He took the view that the primordial moment of its intervention was that of the reproduction of creatures.

In botany, a similar theological orientation played a part in the predilection of the Swedish naturalist Carl Linnaeus (1707–1778) for a classification system based on the systematic analysis of the sexual organs of plants – organs known from the research of the Parisian botanist Sébastien Vaillant (1669–1722). Since then, the Linnaean system has acquired an autonomy beyond the context in which it was initially formed, such that this element of its intellectual genealogy is hardly taught today.

Johann Peter Süssmilch's interest in the ratio of numbers of births of boys and of girls was presaged by the curiosity of one of the first writers on political arithmetic, working before the physico-theological issue developed – John Graunt (1620–1674).

> "[I]n this Parish [of Romsey] there were born 15 *Females* for 16 *Males*, whereas in *London* there were 13 for 14, which shews, that *London*

is somewhat more apt to produce *Males*, then the country. And it is possible, that in some other places there are more *Females* born, then *Males*, which, upon this variation of proportion, I again recommend to the examination of the curious." (Graunt, 1662 [1975, p. 71]).[2]

Half a century later, John Arbuthnot, whose work attempted to demonstrate the action of providence in population figures, took up Graunt's question, applying his knowledge of the calculus of probabilities recently explored by Christiaan Huygens (1629–1695) (Arbuthnot, 1692).

"Among the innumerable imprints of Divine Providence that may be found in the Works of Nature, one, very remarkable to observe, is the exact balance maintained between the number of men and that of women; for, by this means, it is provided that the Species will never be lacking nor perish, since each male can have his female, and of a proportionate age. This equality of males and females is not the effect of chance but of Divine Providence, working to good ends." (Arbuthnot, 1710, p. 186).

His laborious proof was based on the calculation of combinations in a regular game of heads and tails, and took as its starting point the correct idea that, as soon as the coin is tossed more than twice, it becomes less probable that heads and tails will be obtained exactly as many times as all other possible combinations.[3] He took the view that strict equality of the two cases of births, boy and girl, would have to be the result of chance. He observed that known

[2] Graunt – not without some errors, as pointed out by Éric Vilquin, who has recently translated his work into French – found figures for London, from 1629 to 1660, of christenings of 135,324 boys and 125,719 girls (from 1629 to 1664, 157,040 boys and 145,191 girls); and for Romsey (Vilquin, p. 138) from 1569 to 1658, christenings of 3,256 boys and 3,083 girls. He made use of fractional numbers in his approximations, just like his contemporaries. However, the choice of the ratios 14/13 and 16/15 remains debatable. Despite all this, we may reason anachronistically: in London, for 1629–1664, there was a sex ratio of 51.96% and a standard deviation (see Appendix D) of 0.09%; in Romsey, a rate of 51.36% and a standard deviation of 0.63%. The 95% confidence intervals (see Appendix D) around the two ratios are [51.78%; 52.14%] for London, and [50.11%; 52.62%] for Romsey. These two intervals overlap, and the deviation confirmed by Graunt is not relevant, since it must be understood that this means of checking was not available to him.
Translator's Note: "then", which appears twice in this extract, is simply the equivalent of its etymological sibling "than" – "then" was the prevalent form until around 1700. We would like to thank Allison Pollard of the Science Museum Library (Imperial College, University of London) for her meticulous care in transcribing the original quotation from Graunt.

[3] This is because, if one tosses a coin $N = 2p$ times, a gambler who would bet on the two sides appearing an equal number of times would have $(2p)!/p!p!$ for him and $2^{2p} - (2p)!/p!p!$ against him, and this gambler's chances would become gradually smaller, until they reached nil. In terms used today but not in the early 18th century, the equality of the two cases is

counts showed more boys. In its own terms, the physics of the phenomenon (for Arbuthnot, its "nature") differed from the mathematical analysis of combinations of chances, and this difference was a manifestation of the action of providence. But this thinking confused two things that we distinguish more easily today: firstly, the probability of the sex of a birth to come, which he envisaged as a regular game of chance, and secondly, observed births distributed according to sex – a distribution which is not balanced (Hald, 1990; Rohrbasser & Brian, 2005). We know that a game of chance, even a mathematical one, can be unequal. In contrasting the mathematics of an abstract game with the physics of births, Arbuthnot amalgamated two distinctions that we are able to make: the distinction between mathematics and nature on the one hand, and the distinction between the mathematics of an equal game and the mathematics of an unequal game on the other. Could his contemporaries doubt his conclusions? The scholarly literature shows that most of them were induced to follow his calculation, even though gamblers – of whom there were very many in that period – certainly knew that the dice were frequently loaded.

Süssmilch took up the question aired by Arbuthnot, again considering that births could be regarded as a manifestation of the constancy of the action of providence. He brought his extensive material together in a vast work entitled *Die göttliche Ordnung [The Divine Order]* (Süssmilch, 1741). In doing this, he was responding to Graunt's invitation by asserting that male births were generally more numerous than female births, that the ratio of their number appeared to be constant, and that this regularity fell within a notion of divine action which accorded in his eyes as much with natural law as with that emanating from Biblical texts – most particularly, the injunction "increase and multiply" and the principle of monogamy (see Box, p. 6). Thus Süssmilch explored the number of men in order to better understand how divine power accorded with Leibniz's conception of it: omnipresent, always calculating and calculating for the best, with neither nature nor – even less – the physics of human generation having existed before the divine will. Not here the decree of nature, over which the divinity keeps watch even while making alterations to it: probability was the proof of a constant, unique cause, which was said to reveal the divine design. Creatures had to conform to it – although some of them, through reason, were able to throw light on it.

---

the mode of distribution, but – above $N = 2$ – always remains less and less frequent than all the other cases together when N increases.

Süssmilch: The Divine Order in the excess of male births

Since it has previously been proven using sufficient examples that the two sexes are propagated in a very nearly equal balance but with some variation, so that for every 1000 girls born, 1050 boys [51.2% of the total] are always produced, it is clear that more people of the male sex than of the female are meant to live.

But among the births there are not only always more boys than girls; rather, what is truly astonishing is that there is a certain ratio of boys to girls, which is not indeterminate and which does not occur only once or twice by chance but which is constant and almost always the same, as Mr. Derham has carefully observed.

Since it pleases Divine Providence to surround the way into life with so many difficulties, and so many must pass away before they reach the age and situation to serve God and their neighbour properly, God, who in his wisdom makes no choice without sufficient reason, must have had particular intentions for deciding on you to live.

It further follows from this equality, or rather from the proven slight predominance of the male sex, that polygamy cannot take place. [...] And as it is impossible for revealed religion to contradict natural religion, it follows that it cannot permit polygamy.

But wise and Divine Providence forestalls all such things by arranging the propagation of the two sexes in such a way that human multiplication continues in an orderly manner, and that every man finds a helpmate and every woman a husband.

Source: Johann Peter Süssmilch, *Die göttliche Ordnung* (1741, pp. 147, 136, 243, 180 and 133). Translation of this extract from the original German by George Walkden.

2. SCEPTICISM ILLUMINATED

When we reconstruct some elements of the density of discussions on natural philosophy that took place in scholarly Europe from Newton to Kant, pointing out the important place of the physico-theological question in these (Ehrard, 1963; Lagrée, 1991; Loty, 1995), we are not seeking solely to identify traits belonging to the variants of a particular intellectual current, nor even to demonstrate that the formation of economic and social knowledge at the time drew on the panorama of philosophies of the age (after Coumet, 1970; Koyré, 1957; Rohrbasser, 2001; Steinmetz, 2003) – although that alone justifies our in-depth investigation. We also want to show that this scholarly idea occupied a legitimate place within the bounds of new knowledge, even though it was impossible to conceive of a world system that was not under the aegis of a divine power or a superior intelligence, itself conceived in terms drawn from theological dispute (Febvre, 1942; Hahn, 2004), and that the dominant sections of society were accustomed to very high levels of uncertainty in matters of government and economy (Buttay, 2005; Grenier, 1996). For example, it was only after the second half of the 19th century that it became not only legitimate but actually normal to count on stability in economic exchanges and on consistency in the information that accompanied them. This 19th-century movement, involving the taming of chance and the historical formation of collective trust among élites, is today well-known (Hacking, 1990; Porter, 1986, 1995). Perhaps the forms taken by flows of economic information at the turn of the 21st century give us a perspective on doxic adherence to the presupposition of such stability – although in fact the historical duration of this adherence ultimately turned out to be only roughly from about 1850 to the end of the 20th century (Brian, 1996a). In any event, whenever we want to travel back to 1750 through the experience of documented thought, we must take into account the fact that the knowledge and the dominant cultures of those times did not have the same legitimate expedients for reducing the necessary uncertainty of things as the knowledge and cultures of the intervening centuries have had. The result is that legitimate forms of apprehension of this uncertainty should probably be recognized in several features of the culture of the Age of Enlightenment: the craze for games of chance, right up to the salons of the Palace of Versailles (Freundlich, 1989) or even, in clothing, the subtlety of the culture of appearances (Roche, 1989).

"Classical probability" has even been viewed as a kind of scientific culture characteristic of that era (Daston, 1988). But, in this field, the cultural history of the sciences attaches too much importance to their immediate reception. It also tends to iron out some of the disparities in the texts and some of the tensions that structured the production of thoughts that overcame the spirit of their times: in other words, it obscures evidence of the relative autonomy of knowledge in the past. Thus, having sketched the outlines of the probabilistic culture evident among well-read people in the 18th century, we must point out the cleavages between them. In the middle of that century, the success of physico-theology in Northern Europe met with sceptical responses from the direction of the philosophical movement that had already been active for two centuries – particularly in learned societies, and including the most famous of these (Moreau, 2001). Thus, a good number of empiricists with a passion for calculating the number of human beings situated themselves in that intellectual tradition, often placing their researches under the old banner of the Lord Chancellor of England, Francis Bacon (1551–1626) (Bacon, 1620). Thus, the author of *Recherches et considérations sur la population de la France [Research and considerations on the population of France]* (1778), a "person" actually made up of the intendant, Jean Baptiste Antoine Auget de Montyon (1733–1820), and his secretary in charge of calculations, Jean-Baptiste Moheau (1745–1794), wrote:

> "If nature had followed the interests of the propagation of humankind, she would have arranged for more women to be born than men. But the order of production is the opposite, and the male sex would predominate if the causes of destruction [...] did not act more on one sex than on the other. It may be estimated that in France, out of [31] births, there are to be found 16 boys and 15 girls." (p. 130). "Men, who, as we have seen, are born in a greater number than women, lose their superiority from the first year, and their number decreases again in the following years in a stronger proportion." (Moheau, 1778 [1994], p. 166)

Both Süssmilch and Moheau make the same empirical assertions, although in the first they served a theological objective and in the second, the idea of a science of government which, it should be noted, Montyon was trying to press upon Gustavus III of Sweden, the enlightened monarch of a Lutheran kingdom (Brian, 1994b).

It was so easy to call these calculations into doubt, in the name of rigour of observation, that the pamphleteer Louis Sébastien Mercier (1740–1814) was able to call on readers of his *Tableau de Paris [Picture of Paris]* with a plea for more subtlety in appreciating population phenomena.

Mercier was very well-informed about the methods used by those carrying out the calculations, who based their research on perusals of parish registers (in France, the Roman Catholic Church registers), and his response was to point out that an in-depth knowledge of what the 20th-century social sciences have called the *terrain* required them to impose major corrections on their estimates (see Box, p. 9).

Another French example of how calculations could be called into doubt – though more fundamental in scope and scale – is demonstrated in the work of D'Alembert. This was based neither in ruling out the theological thesis, like Moheau and Montyon, nor in hoping for more correct assessments, as Mercier did. D'Alembert took as his aim what, in mathematician's terms, might be called "the metaphysics of the calculus" (Carnot, 1797) – which

Mercier: the objections of an observer of Paris

According to [Monsieur de Buffon], more boys than girls are born in Paris [p. 66]. We estimate [...] that Paris today contains *around nine hundred thousand souls*; and the suburbs, about *two hundred thousand.* Monsieur de Buffon's calculations [seven hundred thousand] and those of Monsieur d'Expilly [six hundred thousand] seem equally faulty.

The *Courier de l'Europe*, in its issue of 3rd July 1781, gave its analysis of the first edition of this work [...]. As the critic's principal objection fell on the idea that I had inflated the population of Paris by taking it to nine hundred thousand souls, I shall respond just to this reprimand by expanding somewhat [...]. *Recherches sur la population de la France*, by Monsieur Moheau, may be applicable to the population in general; but it cannot apply to the capital, because moral causes here surpass physical causes. The comparison of the number of deaths to that of births is not sufficient; the inflow of foreigners forms a class of inhabitants who, so to speak, are neither born nor do they die; the provinces alone pour a crowd of travellers into the city who only pass through, and which is renewed without cease. A public festival sometimes attracts fifty thousand strangers. Paris today has many more inhabitants than it had sixty years ago. Calculations on length of life, which serve as a basis for speculations of this kind, are erroneous as far as Paris is concerned. All the infants who are born there go to be nursed, half die, and the burial registers of the city's parishes are not filled with their names; therefore counts should no longer be based on the register of baptisms, nor on that of deaths.

Source: Louis Sébastien Mercier, *Tableau de Paris* (1775–1782/1994, p. 914 and pp. 987–995).

today might be referred to as epistemology, with no misuse of language other than conscious anachronism (Veyne, 1996).

From the middle of the century onwards, D'Alembert continued to express his perplexity at his era's recourse to the calculus of probabilities when the issue was one of moral sciences or of medicine (for instance 1754, 1767). The first volumes of the *Encyclopédie* gathered together the fragments of his criticism, woven elsewhere throughout the length of his work (see Box, p. 10). His thinking distinguishes the chain of deduction of a mathematical proof in a clearly scientific work from the art of discovering new truths by recourse to the geometers' analysis. It captures the effects of formal abstraction and its faults with regard to what was already known about human phenomena. It marks the limits of physico-theological proofs by making use of the contradiction between faith in the Revelation and

D'Alembert: the doubts of a sceptical geometer

Heads or tails. – [...] To render the solutions to problems relating to games more complete and, so to speak, more usual, it is to be wished that moral considerations would enter into them, relating either to the fortune of the gamblers or to their station or to their situation, even to their strength (when the games are those of commerce), and so on. [...] But since all these considerations are almost impossible to submit to calculation because of the diversity of circumstances, we are obliged to make an abstraction of them, and to resolve the problems mathematically, while at the same time supposing the moral circumstances to be perfectly equal on both sides, or neglecting them totally.

Deduction.– [...] In the matter of sciences, and above all of Logic, deduction is said to be a sequence, a chain of reasoning, by which one arrives at the proof of a proposition [...]. If, in a sequence of propositions, these two were to be found immediately one after the other: *the planets gravitate towards the Sun by inverse reason of the square of the distances, therefore they describe ellipses around the Sun*, this consequence, although correct, could not be said to have been sufficiently deduced, because it is necessary to make the link visible through several intermediate propositions. Thus it could be expressed in this way only in a work whose reader would be supposed to know the link between these two truths through other means. From which it follows in general that in order to judge the good quality of a deduction, the kind of work where it is to be found must be known, and the kind of minds and readers for whom it is intended.

Chance. – [...] As to the manner in which our freedom subsists with eternal Providence [and] with the immutable laws to which all beings are subject, it is an incomprehensible secret from us, of which it has not pleased the Creator to reveal the knowledge to us; but what is perhaps no less incomprehensible is the temerity with which some who believe themselves or are said to be wise have undertaken to explain and to reconcile such mysteries. In vain does the revelation assure us that this abyss is impenetrable; proud philosophy has undertaken to plumb it, and has only lost itself there [...]. The true philosopher is neither Thomist, nor Molinist, nor Congruist; he recognizes and sees everywhere God's sovereign power; he avows that humankind is free, and keeps silent on what he cannot understand.

Source: Jean Le Rond D'Alembert, *Encyclopédie*, vols IV and VII, (1754–1757).

the systemic mind in philosophy. D'Alembert did not expect an incorrect analogy between games of chance and human affairs, especially when argued in the name of mathematics, to provide proof of the action of divine providence; the physico-theologians' conclusions had no basis in any such proof; and the important thing for him was, above all, to establish rigorous observations of the phenomena in question. Showing his affinity with other contemporary empiricist tendencies (Gillispie, 1980; Roche, 1993), he took the view that it was better, for example, to organize a Royal Society of Medicine to oversee the work of observers of inoculation against smallpox, than to speculate on its efficacy using a poorly-based calculus of probabilities (Brian, 1996b; D'Alembert, 1767).

Was D'Alembert thinking of Arbuthnot and Süssmilch? We may be sure that he read carefully the *Philosophical Transactions* of the Royal Society of London and the *Mémoires de l'Académie de Berlin*, where the two physico-theologians had published their writings. He was completely familiar with the research being conducted in the learned society formed in Prussia by Frederick II, and its impact on philosophy. But, like most of his colleagues in the Royal Academy of Sciences of Paris, he adhered to one of the characteristic features of this institution. From its foundation in 1666, then from its renewal in 1699, issues of religion, and therefore of theology, were excluded through the combined effect of the religious policy of Louis XIV's Absolutism and a determination to keep one of the main elements of conflict between 17th-century Parisian scholars in the background (Demeulenaere-Douyère & Brian, 2002). But a century later, in Berlin, for example, theologians and mathematicians rubbed shoulders, even occasionally collaborated – as did Süssmilch himself, with Leonhard Euler, D'Alembert's chief rival in his time (Süssmilch, 1741).

Thus, in total conformity with the social context of the Parisian institution and in intense Europe-wide competition with the mathematicians Leonhard Euler (1707–1783) and Daniel Bernoulli (1700–1782), D'Alembert mobilized the sceptical tradition, calling into doubt the calculus of probabilities of his day. By doing so, he attracted the scorn of a large part of mathematical Europe – even of posterity – which took the expression of his doubts for a profound incapacity to grasp the significance of a very fashionable calculation.

Nor have the subsequent attentions of philosophy favoured the fact that D'Alembert's expressions of scepticism were fragmentary, conveying that this was a matter of conscience and displaying the caution of an unrivalled mathematician. Posterity has preferred to look to his Scottish contemporary and correspondent, David Hume (1711–1776), for clarification of questions of induction and numerical inference (Hume, 1748 [1988]). Thus for two-and-a-half centuries, the paths of mathematicians, philosophers and those conducting empirical calculations have crossed and recrossed the territory of induction, where each has gone forward, indifferent to the labyrinth of translations from one learned specialism to another, from one era to another, from one language to another (especially between English, French and German). Therefore it can be seen that the real difficulty confronting us here does not lie in how to divide the number of births of one sex into the total number of cases observed, but in being able to trace a coherent, rigorous course across three centuries of phenomena and knowledge in a relevant and effective manner.

3. THE BEGINNINGS OF THE ANALYTICAL THEORY OF PROBABILITIES

Laplace and Condorcet, both disciples of D'Alembert, and amongst the most active European mathematicians in the last decades of the 18th century, had to take stock of that part of the scientific legacy left by their elder. This was the starting-point for an investigation – now fairly well-documented – into the beginnings of the "analytical theory of probabilities", an expression that Laplace was to use as the title of the book that, from 1812 onwards, would be the source for his mathematical successors (Brian, 1994a; Bru & Crépel in Condorcet, 1994; Bru, 2003; Gillispie, 1997, 2004; Hahn, 2004).

Coming as they did on the eve of the French Revolution, there was an intrinsic connection between the early genesis of the formation of the calculus of probabilities in the style of Laplace – a calculation which would govern a later development in the sciences frequently described as "the probabilistic revolution" (Krüger, 1987–1989) – and that of the procedures in administrative statistics that would characterize 19th-century nation-state building. Two processes had a profound impact on both the history of mathematics and the history of administration: firstly, the increasing autonomy of the state from the absolute monarchy and secondly, the great success of the printed book, especially in the areas of moral and political sciences and of economics (Brian, 1994a). Moreover, we should take into account the importance of two critical tensions among the scholars concerned. The first was institutional in nature. The Royal Academy of Sciences of Paris, as reformed in 1699, was reaching the limits of its usefulness in the absolutist government machine. Its scholars, guarantors of the useful sciences, were rendered somewhat superfluous by the flourishing of the bookshop. The second, epistemological in nature, brought into play the significant application of forms of mathematical thinking created during the development of differential and integral calculus throughout the 18th century – which is where

D'Alembert's scepticism found its place. His doubts about the calculus of probabilities were in fact an expression of the art of reasoning – the art that gave rise to the success of this exceptional geometer in the mid-18th century.

The result was that the younger generation of mathematicians was forced to move forward by strengthening the learned tradition that had fostered them. The epistemological crisis and the institutional crisis reached the same resolution in the scientific and political conditions of the end of the *Ancien régime* (Brian, 1994a; Gillispie, 1980). From the early 1770s, Condorcet and Laplace sought to renew the calculus of probabilities. The former was D'Alembert's disciple most in the public eye: he soon became the Permanent Secretary of the Royal Academy of Sciences – that is, one of the most important personages dealing with questions of publication and censorship of scientific research in that company (McClellan, 2003). The latter was at that time a figure newly-arrived on the Paris academic scene from the provinces: a prolific, even dazzling, young mathematician, he rivalled Condorcet, who was his elder by some years (Hahn, 2004).

Following criticisms of D'Alembert, and because these brought into play the art of reasoning by mathematical analysis, both Condorcet and Laplace found that this process of renewal involved moving away from reasoning in the style of Blaise Pascal (1623–1662), Christiaan Huygens (1629–1695) or Jakob Bernoulli (1654–1705) – that is, posing an analogy between the calculation of combinations in games of chance and the measure of any probability. Simply counting favourable cases and possible cases was a convenience that would have to be revised. Using the integral calculus, they perfected a method for measuring particular causes through the cause of events that had taken place – this was the line followed by Laplace – or for measuring the *motive to believe* (Pearson, 1978) in a future event, once comparable past observations were taken into account – which was Condorcet's line. Both tackled this research by embarking, in the 1770s and 1780s, on an extensive Europe-wide scholarly literature. Everything suggests that one of the first readers to whom they submitted their essays, Turgot himself (1727–1781) – not yet a minister but already an intendant and one of those in *Encyclopédiste* circles who was most familiar with English philosophy – drew their attention to the Reverend Thomas Bayes' calculation method (1702–1761), published posthumously some years earlier by the Dissenting minister Dr Richard Price (1723–1791) (Bayes, 1764). A calculation mechanism strictly comparable to Bayes' was used by Laplace in his early papers on the calculus of probabilities, accompanied by elements of the theory of a new class of function – the extensive development of integral calculus, made necessary by the complexity of the equations to which it led. The calculation thus perfected allowed the comparison of the favourable to the possible to be taken up again, no longer reasoning on the basis of crude counts, but of differentials constructed in such a way as to grasp the object of the analysis, and then integrated according to the whole range possible for its variations (Brian, 1994a).

Boasting of the scope and significance of his method, Laplace immediately applied it to the problem of the proportion of boys and girls at birth. He was thus able to distinguish the "greater or less facility of births of boys relatively to those of girls" – that is, the cause that, in his eyes, made the birth of boys more or less probable – from the observed counts of births of boys and girls – that is, events that had taken place. The calculation aimed to grasp causes through events: more precisely, not to highlight a primary cause as the theologians had hoped, but to analyse and compare various causes whose measurement would be enabled by the new calculus (see Box below).

Laplace: how to analyse causes

I give [...] the solution to some interesting problems in the Natural History of Humankind such as that of the greater or less facility of births of boys relatively to those of girls in different climates. [...] For this delicate research, much bigger numbers [than four or five hundred] must be used, given especially the slight difference that exists between the facility of births of boys and of girls; and it is only when one is well assured that the number of births observed in any place indicates with very great probability that births of boys there are less possible than births of girls, that it will be permissible to research the cause of that phenomenon. [...] The probability that, in Paris in one year, births of boys will not be in a greater number than births of girls, is [...] less than 1/259. [...] The probability that births of boys in London will not surpass that of girls, in a given year, is therefore a little less than 1/12416. [...] This phenomenon is, as we see, much less probable in London than in Paris, which comes of the fact that in the first of these cities, the ratio of births of boys to births of girls is more considerable. The odds [are] more than four hundred thousand to one that births of boys take place with greater facility in London than in Paris; thus it can be regarded as very probable that there exists in the first of these two cities a cause, more than in the second, that facilitates births of boys there and which depends either on the climate or on the nourishment and the customs.

Source: Laplace, *Mémoire sur les probabilités* (1778), pp. 228, 274, 282–283, 312–313.

Condorcet's commentary on Laplace

Thus Monsieur de la Place finds that there is a very great probability, almost equivalent to a moral certainty that the excess of the number of births of boys has a physical cause for Paris, that there are odds of 259 to 1 that in the next year the number of girls will not exceed that of boys [...], and that the certainty that this effect has a regular physical cause is incomparably greater for London than for Paris.

Source: Condorcet, *Histoire de l'Académie des sciences* (1778), pp. 44–45.

Thus, he gives figures for births in Paris between 1745 and 1770, of 251,527 boys and 241,945 girls, which is a ratio of 105 to 101 (a proportion of 50.97% boys); while for London the numbers from 1664 to 1757 were 737,629 boys and 698,958 girls, which is a ratio of 19 to 18 (a proportion of 51.35% boys) (Laplace, 1778). Rather than seeing the difference between these two proportions as an index that can be interpreted directly, he wonders what the odds are that each of these two series of observations is caused by the same level of "facility" of boys being born. In other (more up-to-date) words, he wonders at what point these figures could be obtained in the situation where a newborn infant had the same probability of being a boy in Paris and in London. In using the word "facility" to indicate the cause of sex at birth and the word "probability" to judge the odds on each hypothesis, Laplace is making a pertinent distinction between what we conceive of as, firstly, the law of probability that might account for each occurrence of the phenomenon and, secondly, the probability of observing such an empirical result.[4] However, although he is in

[4] Condorcet, as we shall see, was to clarify this distinction some years later. Today we can understand that "facility" (as used by Laplace) relates to the law of probability of the random variable that characterizes the phenomenon, while "probability" (still in Laplace's sense) is peculiar to the estimator of the parameter of that law, constructed on its being repeatedly realized. But from the point of view of statistical calculation and of the history of sciences, it is important not to rush ahead, since another extrapolation of Bayes' formula in the calculation that today we call "Bayesian" poses – and here we conform to Jakob Bernoulli's vocabulary (1713, pp. 224–225) – a priori probability and a posteriori probability (this time the sequence of the two different times is significant). It should be stated that the same formula as used by the Reverend Bayes shaped two distinct concepts, both present today in the panoply of the statistician. It is not the job either of historians or of epistemologists to unify the two, but only to state their meeting points. It is the responsibility of mathematicians to explore their predecessors' texts with a view to finding new routes out of old, misunderstood passages. Thus certain of Condorcet's passages may be understood in parametric terms (e.g., in Condorcet, 1785) and others in Bayesian terms (e.g., in Condorcet, 1784–1787). See, on this subject,

the right here, it is clear that he revised his writings on probabilities before his publications in the 1810s (Hahn, 2004). In order to establish a theory, it is not enough to envisage and carry out a coherent calculation and to be almost its only careful reader. This acknowledged epistemological fact is validated by a historical fact: it took Laplace thirty years to reach the point where he could express his conception of the calculus of probabilities in a way that he judged acceptable and that posterity recognized as such for a century.

For his part, Condorcet did understand the calculus; but it must be observed that there was a strong tension between the two men. However, this did not extend to controversy – because before 1789, they alone were capable of grasping the object of their researches, and only a few people came to do so later. Thus Condorcet understood that Laplace's particular contribution avoided an element that was of primary importance to the metaphysical debate of his time: that observations were necessarily past and that the method established a measure of a *motive to believe* – that is, some degree of certainty about things to come. Their two conceptions of an analytical theory of the calculus of probabilities thus met in a dialogue of conflict, woven throughout the *Mémoires de l'Académie royale des sciences* (notably Condorcet, 1781, 1784–1787; Laplace, 1778, 1785, 1786a, 1786b).

From the 1770s onwards, Condorcet – like Laplace later, no doubt under his influence – realized that they were confronting the Humean challenge to numerical induction (Baker, 1975; Hahn, 2004). After D'Alembert, they rejected the analogy with games of chance, used since the 17th century by the founders of the calculus of probabilities – an analogy so perilous to any areas bordering on theology – but without abandoning either the application of the new calculus to those games or the comparison of a measured probability with that of a supposedly known game. This reform of numerical induction acted as a springboard for the empirical analysis of causality, and has become continually stronger ever since (Fagot-Largeault, 1989; Hacking, 1975, 2001). With Laplace and Condorcet, the distinction between probability understood as the cause of the phenomenon, case by case (it is as if the sex of the newborn baby were more often boy than

---

Bru and Crépel's comments in Condorcet (1994), Pearson (1928, 1978) and Todhunter (1865). More generally, useful elements for the assessment of probabilist and statistical research can be found in Grattan-Guinness (2005); Hald (1990, 1998), Heyde and Seneta (2001), Krüger (1987, 1989) and Stigler (1986, 1999). In the edited works on this list, Ivo Schneider's articles are particularly relevant. See also the specialized journal, the *Electronic Journal for the History of Probability and Statistics* (www.jehps.net).

girl, in the measure of 0.512 to 0.488), and the frequency of observations (51.2% male births were counted) became sufficiently elaborate mathematically – that is, relieved of recourse to any exogenous argument – that later mathematicians were able to develop the calculus of probabilities and its applications without having to return to the physico-theological debates of the 18th century. But from its earliest days, as the radical differences between Laplace's preoccupations and those of Condorcet (and even those of Bayes) show, this new calculation and the relationships that it maintained with counting operations did not give rise to a univocal doctrine. It was – to use Condorcet's expression – a "technical method" (Condorcet, 2004), and today's specialists in that method are still exploring the diversity of ways of thinking that it can call on (Hacking, 2001; Hald, 1990, 1998; Pearson, 1978; Stigler, 1986, 1999).

4. DOES HUMANKIND GAMBLE ON PROBABILITIES?

D'Alembert pleaded the cause for more observations. Süssmilch himself prescribed instructions with a view to "the good ordering of parish registers" (Süssmilch, 1741 [1998, pp. 297–304]). Condorcet used the commentaries he was required to give – in his capacity as Permanent Secretary – in the columns of the volumes published by the Royal Academy of Sciences of Paris, to set out his wish to improve records of observations on births and deaths (Condorcet, 1782). A manuscript in his hand has also been found, which seems to have been compiled in the same period, entitled *Arithmétique politique ou application des mathématiques aux sciences économiques [Political arithmetic or the application of mathematics to the economic sciences]*, where he renewed the same recommendations, stipulating that it was necessary to register "the number of births, distinguishing males from females", and citing a work that reproduced and identified Süssmilch's tables and those of others who had undertaken such calculations.[5]

It should be made clear that, from 1772, the French government had been collecting the parish registers of births, marriages and deaths for the

[5] This text was published by Bru and Crépel in their edition of Condorcet (1994, pp. 338–341). They observe that the "very good work on the calculus of probabilities" mentioned by Condorcet, which he attributes to Fontana, is in fact Roberto Gaeta's thesis, *La Dottrina degli azzardi*, Milan, Galeazzi, 1776, supervised by Gregorio Fontana, which was constituted from a translation of Abraham de Moivre's treatise, *The Doctrine of chances*, London, 1756, and a series of extracts from other authors, including Süssmilch.

whole kingdom in Paris, but that recording the distinction between the sexes had been abandoned under Necker's Ministry in the late 1770s. However, the intendancies – then the principal machine of government activity in the provinces – had devoted considerable energy to it. As a result, in 1785, when the Academy of Sciences found the means to procure the results of the compilations produced by the administration, Laplace abandoned the idea of examining further the probability of the ratio of the sexes at birth, in favour of applying his new calculus to the probability of the ratio of the number of inhabitants to the number of births registered in the parishes – a relative ratio that was then of much greater use to political arithmeticians and economists in evaluating territorial wealth (Brian, 1994a).

In the same academic volume (the one for 1783, which appeared in 1786), the conjunction of the deaths of the two greatest mathematicians of the 18th century, Euler and D'Alembert, with that of the astronomer and Secretary of the Stockholm Academy of Sciences, Pehr Vilhelm Wargentin (1717–1783), gave Condorcet the opportunity to deliver three eulogies where, in passing, he assessed the work of the previous generation on the question of probabilities and enumeration. He pointed out Euler's work on life assurance and mortality tables, but not his collaboration with Süssmilch. A few pages further on, discussing Wargentin, he praised the "commission charged with gathering all the details relating to the population of Sweden". In his eyes, the sagacity of his opposite number in the Northern Lutheran monarchy related to "the art of deducing their general results from observations". He was in effect pleading for prudence with regard to physico-theology and for the validity of his own activities in Paris: "it was believed, in Sweden, that a skilled mathematician could, when it came to pronouncing on the results of calculations, sit alongside the members of the Administration". Recalling the judgments passed on D'Alembert also gave him an opportunity for clarification and correction.

> "He was accused of attaching little importance to the physical sciences [in the sense of the natural sciences], and this accusation was unjust; he scorned only those systems whose proofs are confined to showing that the absolute impossibility of them has not yet been rigorously demonstrated; those uncertain general ideas that are announced as grand plans; those explanations relying on vague thinking, which could at the very most lead to slight probabilities, and finally that abuse of scientific language which sometimes changes into a science of words which ought to be only a science of facts and of calculations. It may be believed only that he pushed rigour too far, for even if these hypotheses, these views,

these explanations do not form a true science, they serve to multiply experiments, observations, to show them in their different guises; they guide us in our research, they open the way to discoveries, and seem to be the dawn of the day that the centuries to follow us may hope to enjoy." (Condorcet, 1786).

During the 1780s, Condorcet, although he attached great importance to the use of population counts in matters of government, remained no less dissatisfied with the state of the art of conjecture, even when it was improved by his colleague Laplace. Like D'Alembert, he envisaged the application of mathematical analysis to probabilities first and foremost from the point of view of a metaphysics of calculation, and this allowed him to envisage a horizon of expectation for the sciences to come. Of course, he granted the young Laplace his technical advances – but not his attempts at philosophical discussions. In his eyes, the stumbling-block for the new methods, as for the old, was the question of the forward-looking nature of observations and the backward-looking nature of the motive to believe. This perplexity explains the importance, in his *Mémoire sur les probabilities [Paper on probabilities]*, which appeared in the Academy's volumes between 1784 and 1787, of his *Réflexions sur la méthode de déterminer la probabilité des évènements futurs d'après l'observation des évènements passés [Thoughts on the method of determining the probability of future events according to the observation of past events]* (Condorcet, 1786), where he distinguished his position from his colleague's in *Mémoire sur la probabilité des causes par les évènements [Paper on the probablity of causes by events]* (Laplace, 1774). He took care on this occasion to distinguish the two moments of the analysis – past and future. Moreover, he stated precisely that first of all a "law" should be laid down that was capable of governing the probability of every occurrence of the phenomenon and that then a probability of another order should be measured (for example, the probability of observing such an event in the future). This was not all. He distinguished, in passing, between the hypothesis that past events would obey this law exactly and the hypothesis that this would only nearly and on average be the case. Finally, he indicated that either it should be supposed that "all events [are] independent" or else recourse should be made to more complex hypotheses, taking into account the chronological order of past events. His thinking aimed to clarify abstract hypotheses, calling for a realistic, rational application of the analysis of the calculus of probabilities. The article brought together in a single concept the measure of the motive for believing that there is a constant law of nature, the weakening of this probability with time, and the distinction between the repetition of rigorous observations (this was the hypothesis of the effect of a single unknown law properly observed several times) and the multiplication of imprecise observations (the hypothesis of different but fairly similar laws).

The last paragraph dealt with two sequences of past observations, S and S', of two conflicting events A and N, where any reader familiar with the context could easily recognize the abstraction of two lists of birth records (series S and series S') divided into boys and girls (A and N, or the reverse). The numbers of As and Ns observed were m and n for S, and m' and n' for S'. Condorcet proposed to measure the probability of finding, in a future series that would correspond to a group of hypotheses finally made explicit, the As and the Ns in number p and q.

His conception of the calculus of probabilities allowed Condorcet to outline an analytical framework that enabled him to say with precision that "a natural fact observed for single time, provided that it has been well observed and analysed [...] may be regarded with very great probability as a constant fact" and to take the view that, if a law is established today as almost certain, "there may be a more complicated constant law which for a time seems the same to our eyes as that which was first established and which then deviates from it significantly" (Condorcet, 1786, pp. 548, 551). The historical time, mathematical abstraction and the degree of certainty that reason may expect here and now were combined in a single epistemology (of course, the word is anachronistic), a metaphysics of calculation or a philosophy based in analysis (this time, these terms are appropriate to the era).

This theory of knowledge, fostered by contemporary European discussions, answered some of the questions known to be the most profound in Enlightenment philosophy.

> "As far as causality is concerned, for example, it cannot be denied that Malebranche [1638–1715], placing every cause in the Christian God, thinks in accordance with his religion and his faith. Berkeley [1685–1753], seeing in the world the language that God speaks to us, finds in his fashion the Biblical image of Genesis. Discovering the source of causality in human nature, Hume [1711–1776] conforms to the trends of an atheism fashionable in his times. And the Kantian idea [Kant, 1724–1804] of legislation is not without relationship to his pietist inspiration. Therefore it is not surprising that the systems of these philosophers are opposed, depending on their conclusions, and depending on their methods [...]. The only more remarkable thing to be stated is that, freed of the prejudices that objectivize them, stripped of the language that

systematizes them, the theories of Malebranche, Berkeley, Hume, and Kant are revealed as plainly identical. For they establish with rigour that no objective, temporal sequence can offer us what we understand by cause." (Alquié, 1957, p. 13).

From a historical point of view, we know that Condorcet could have had Malebranche and Hume in mind (we should also add Locke), that he was not familiar with Berkeley's works, and that he did not know Kant's philosophy – but that he responded in various ways to much contemporary thinking on the relationship between historical time and the abstraction of calculations (Cléro, 2004; Coumet, 1970; Crampe-Casnabet, 1985; Daston, 1988; Loty, 1995). His analytical conception of the calculus of probabilities, delivered in fragments, was in any event a more closed metaphysics than that of Laplace.

Once having pinpointed this epistemology, it is easy to link together Condorcet's two best known works today – even though commentaries on them seem to find it necessary to stray into areas of science foreign to one another: the *Essai sur l'application de l'analyse à la probabilité des décisions rendues à la pluralité des voix [Essay on the application of analysis of the probability of decisions made by a plurality of votes]* (1785) and the *Esquisse d'un tableau historique des progrès de l'esprit humain [Sketch for a history of the progress of the human mind]* (1795 [2004, pp. 233–459]). The first proposed an application of the new calculus of probabilities according to a set of complex hypotheses on greater or less independence between votes cast in the jury voting process. The second, which was Condorcet's last work in his scientific programme, gives the prospectus for an enquiry into the history of knowledge and of human consciousness – these are its first nine "Epochs" – and an extrapolation of what he thought it was reasonable to believe about the future of humanity – this is the Tenth "Epoch". The whole work gave a broad assessment of the conditions which, according to this analysis, could – or even might possibly – favour or hinder the extension of the horizons of humankind in knowledge and science.

We know that the writing of the *Tableau historique* was never completed. When, a short time before his death, Condorcet left the Paris refuge where he had hidden during the Terror, he left behind numerous manuscripts, some fragments of which were published from time to time during the 19th and 20th centuries. Today we have a complete edition, which allows us to measure at what point the logics of the style of narration peculiar to the genre of the prospectus overly imprinted a teleological movement on the *Esquisse*, making a lasting mark on the philosophy of the history of knowledge. The main body of the *Tableau historique* is not so simplistic. It teems with more subtle critical discussions – which, for all that, is not to say we should adhere to the philosophy of history that emerges newly from it (Condorcet, 2004).

One of these developed discussions deals directly with the issue of the sexes at birth. In opposition to intellectual censure – whose strength he attributes to religious superstition – Condorcet looks at this question without ever calling on divine power for help; he even wonders whether it is possible for humankind to alter the probability of the sexes at birth: that is, to exert an influence, with the aid of knowledge that is only probable, on what we now call sex ratio at birth (see Box, p. 23).

In making any assessment of the 18th century, we must note that analysis of the proportion of the sexes at birth was one of the empirical terrains where metaphysical and mathematical debates were liveliest. By the end of the century, there were four conceivable positions. The first, predominant in the specialist literature, was that of physico-theology: the continuously-proven regularity of the excess of males births was there held to be the manifestation of a unique, constant cause, and the mark of divine providence. A second, conforming to the scepticism that was shared by a number of scholars and observers of the Enlightenment, was expressed in its full mathematical rigour by D'Alembert – risking the scorn of the scholarly Europe of his time: it consisted of calling into doubt the validity of applying the calculus of probabilities in the style of Pascal or the Bernoullis to human affairs. Two other positions can be located within the narrow circle of geometers at the Paris Academy of Sciences: firstly, that of Laplace, who set out to measure the probability of the action of various particular causes in the diversity of known proportions of the sexes at birth; and finally, that of Condorcet who, careful to refine the hypotheses necessary for applying the calculus of probabilities and wishing to free the issue from superstitious prejudices, arrived at the idea that human beings could have some effect on the probability of the birth of one sex or the other.

Although these four views can be shown to have existed and are linked together by identifiable historical conditions, it would be false to say that they cover the whole spectrum of knowledge common around 1790. Laplace's papers were then almost incomprehensible, whoever the reader. His intellectual rivalry with Condorcet was perceived only by some scholars,

Condorcet: an ability of humankind

Effects on the moral and political state of humankind of some physical discoveries like the means of producing male or female children, as one chooses, with a certain probability; of producing children without the union of the mother with any man; etc., which may have results acting for or against the continued perfecting of humankind.

Rarely have philosophers directed their assured gaze upon those objects located between disgust and ridicule, where both hypocrisy and scandal must be avoided. Christian superstition, preoccupied with the chimeras of a mystical purity, has led us into the habit of attaching ideas to these physical actions, which for good or ill may be applied only to their moral consequences [...] But it is time to rise above the hypocrisy of customs and the hypocrisy of style, showing objects in their true light and under their own colours. [...]

In the hypothetical case that one had a means of determining the sex of children at will, at least up to a certain point, would there arise a significant difference in the number of individuals of each sex instead of the almost total equality that exists today, and which then would be the most numerous, and what might be the effects of this disproportion on the social order? [...] Does not each sex have an interest in multiplying the sex that it is not? [...]

The only result of this discovery [...] would be a means of re-establishing an almost total equilibrium either in the small portions where nature alone would not have established it or in the extraordinary circumstances where she would have deviated from it. There would be a real danger if one sex had a pressing interest, opposed to that of the other sex, in augmenting the proportion of the number for itself, but that interest does not exist, and consequently force could not preserve a disproportion contrary to the general progress of humankind once chance had established it.

Source: Condorcet (1793–1794 [2004]), *Fragment 10*, pp. 923–937.

and the most advanced aspects of the analytical theory of probabilities were not necessarily clear to them. The main part of Condorcet's "Fragment 10", moreover, was not published until the early 20th century (Cahen, 1922). Among young people who were being trained in the sciences at the very end of the 18th century, the view would possibly – if at all – have been taken that the mathematical analysis of the excess of male births was a particularly delicate issue, comment on which was reserved to the sole surviving elder: Laplace (Condorcet died in 1794). It is not impossible that some of these younger scientists – such as Joseph Fourier (1768–1830) – had the opportunity to consult Condorcet's papers, but nothing has come down to us today to support that conjecture. Whatever the case, the traces that historians of the sciences – for all that they might very much like to follow them – still struggle to make out more than two centuries later must lie hidden in oral transmission and in the leisure that all then had for careful reading of the *Mémoires de l'Académie*.[6]

However, one testimony to the legitimate state of knowledge on the subject of sex ratio at birth in France at the very end of the 18th century does still exist. This is the tenth lesson, given on 21 Floreal, Year III (10 May 1795) by Laplace himself to the newest students of the *École normale* (Laplace, 1795 [1992]). In this, he summarized his pre-Revolutionary results. Going further than either Süssmilch or D'Alembert, but ignoring Condorcet's speculation, he affirmed the existence of a "general law of nature", according to which there were always more boys than girls at birth, although their proportion was not necessarily equal everywhere. He also wondered about the cases "of diverse species of animals and plants". The analytical calculus of probabilities, in his eyes, offered the advantage of rigorously establishing the extent of the observations that would have to be mobilized – he indicated "several million births" – in order to "acquire a great probability of the existence of causes [that these observations] seem to indicate, and distinguish them from those variations that chance alone brings in the sequence of equally possible events" [p. 135]. From then on, this became the central issue in the analysis of the proportion of the sexes at birth: chance alone, or constant causes that could be established through calculation? Oscillation around the schema *chance alone or constant causes* – a schema sustained by research and commentary since the early 19th century – is nowadays a given of collective memory, whether academic or lay. Therefore, if we want to free ourselves from this dichotomy, we should be ready to reiterate Durkheim's gesture of rejecting *praenotiones*

[6] We can imagine the state of mind of this new generation – for whom the analytical reform of the calculus of probabilities by Laplace or Condorcet was clearly an intrinsic part of the flourishing of the differential and integral calculus of the late 18th century – on reading Carnot's *Métaphysique du calcul infinitésimal*, where the term "indeterminate" was understood as used by Descartes but not with the idea of what was later called "Laplacian determinism". Thus, Carnot said of the mathematician Lagrange (1736–1813) "I have heard this profound thinker say several times that the true secret of his analysis lay in the art of grasping the diverse degrees of indetermination to which quantity is susceptible" (Carnot, 1797, Vol. 2, p. 67). Condorcet distinguished two orders of indetermination: the first related to the treatment of approximation through differential calculus, and the second to the measure of the degree of uncertainty that any phenomenon involves. The distinction of this second order characterized, in his eyes, the new calculus of probabilities (Brian, 1994a).

(Bacon, 1620; Durkheim, 1895) – although in this particular case the preconception arises from the science itself. In other words, we should repeat the gesture of breaking with spontaneous sociology (Bourdieu, Chamboredon, & Passeron, 1968), although this time it is not a form of sociology, but a form of spontaneous epistemology. In doing so, we may expect a new syntax to emerge, combining chance, calculation and social fact.

Chapter 2

PHYSIOLOGY, PROBABILITIES AND STATISTICS (1795–1830)

1. ADMINISTRATIVE COUNTING AND ANALYSIS OF CHANCES

The layout of the first map of the *départements* of France, made in 1790, was ruled by the geometrical spirit: it was made up of entirely equal squares (Ozouf-Marignier, 1989). Similarly, register office reform was a manifestation of the scientific spirit of the 18th century, encouraged by the political and administrative opportunity offered by the French Revolution (Aberdam, 2004). The lack of simplicity of registration and its minimal uniformity, the absence of systematic registration and the weakness of its wording were all deplored in the literature of economics and political arithmetic in the final years of the *Ancien régime*. From the Revolution onwards, Laplace's and Condorcet's papers on population, which appeared in the final volumes of the old Royal Academy of Sciences, served as models for innovations, alongside those by a few other scholars. Laplace himself was Minister of the Interior for a few weeks under the Empire – but this is of only passing interest, since the office proved so foreign to him.

Around 1780, Condorcet, always so abstract and so distanced from the administrative reality of the *Ancien régime*, mused: "The manner of arrival at obtaining an exact enumeration of a country and of burial lists made with precision depends uniquely upon administration" (Condorcet, 1994, p. 339). Thirty years later, this view of the necessary spirit became a legitimate design, governing both the ideal for implementing administrative organization and the aim of scholarly research.

From a sociological point of view, this was the time of a new division of labour between academia and administration. But what are the effects of

E. Brian and M. Jaisson, The Descent of Human Sex Ratio at Birth, 27–57.

reading this period in such terms? In this respect, it is illuminating to revisit certain aspects of Durkheim's thesis.[1]

> "Each science has, so to speak, a soul which lives in the conscience of scholars. Only a part of this soul assumes sensible bodily form. The formulas which express it, being general, are easily transmitted. But such is not the case with this other part of science which no symbol translates without. Here, all is personal and must be acquired through personal experience. To take part in it, one must put oneself to work and place oneself before the facts." (Durkheim, 1893; translation from Simpson in Durkheim, 1933, pp. 362–363)

Here, he was already – and very few sociologists of the sciences have noted it since then – distinguishing between three essential components of academic activity: firstly, the paraphernalia of formulas and methods, which can fairly easily cross the centuries; secondly, the spirit of the scientist, which is not determined in formal pathways; and finally, the craft involved, the confrontation with the things of the sciences, without which the scientific mind cannot arrive at the formulas. A change in the social division of scientific labour does not affect these three aspects in the same way.

This is because formal pathways and material supports, although they can move from one situation to another fairly easily, are nevertheless reappropriated differently in new conditions: this was the case with Laplace's early calculations on the subject of probabilities, which he was to spend thirty years reformulating in order to hand them on to his successors (Laplace, 1812). Academic meaning or know-how – in other words, habitus – may be found there again or not: everything depends on the nature of the changes. The scholar then has to intensify his explanations, and express his own workings himself. This is the reason behind the development of the introduction to his *Théorie analytique des probabilities [Analytical theory of probabilities]* (1812–1820), which was rapidly to become the *Essai philosophique sur les probabilités [Philosophical essay on probabilities]* (1814–1825).

[1] Durkheim's thesis, *De la Division du travail social* (1893), allows us to grasp some of the conceptual implications of an analysis in terms of the social division of labour. But in his mind, such a process of division was directed towards historical progress. However, it is not necessary to share this teleological perspective and the philosophy of history that it entails, in order to think in terms of the history of the social division of production of symbolic instruments (Bourdieu, 1977; Brian, 1994a, 2001a).

By the end of the transformation of the division of labour in the production of symbolic instruments, concrete conditions had themselves changed radically. This change has been recorded by historiography over the last three decades, following several studies on early 19th-century methods of enquiry (Aberdam, 2004; Bourguet, 1989; Chappey, 2002; Guégan, 1991; Margairaz, 2005; Ozouf-Marignier, 1997; Pansini, 2002; Perrot, 1977).

Following this transformation in the division of labour between science and administration, the various protagonists preserved distinct fragments of collective memory, which first became uncoupled during the transformation process and then combined in different ways. In these conditions, the phenomenon that was being studied over a very long period of time – the proportion of the sexes at birth – cannot be grasped unless these shifts in collective memory are disentangled. Social science investigation nowadays, for all that it aims to detach itself from the most superficial explorations, must therefore mobilize, on the one hand, intellectual history and the history of the sciences and, on the other hand, the study of social and economic phenomena, barring detours into any sort of positivist false pretences or relativist illusion.

Revolutionary and post-Revolutionary institutions (in particular, the *École normale* and the *École polytechnique*) and the establishments of science, administration and the armed forces were likewise fashioned by scholars or even by politicians accustomed to the most recent sciences, which they had come to know during that period of tremendous opportunities – although it is a given that they had to correct their initial views as they went along. The Office of Longitudes and the First Class of the Institute (successor to the old Royal Academy of Sciences), both created in 1795, regularly recorded the works of these same scholars (Gillispie, 2004; Heilbron, 1990).

As far as births in particular were concerned, during the decades just before the Revolution in France, the Catholic Church was recording the main sacraments, and so obtaining fairly complete lists of a large proportion of christenings; later, the provincial intendants and the Controller-General of Finances made annual collections of parish records and arranged them in tables, or even matched them to the geodesic map of the kingdom, in order to estimate fluctuations in local territorial wealth through variations in births and deaths. In exceptional circumstances, some of these tables were sent to the Royal Academy of Sciences, where Condorcet, Laplace and a colleague published extracts from them, in the hope of reaching the standards of the equivalent Swedish Company of Scholars and of providing an example of rigorous compilation, more or less justified by the new calculus of probabilities, for use by economists and others making calculations (Brian, 1994a).

Surveys and calculations proliferated in all spheres of administration and government from the start of the Revolution. Numerous recent studies have shown the difficulty – even to the point of incoherence – of organizing specialized research, most often carried out under the leadership of well-trained scholars, if not academicians. The result was that, by about 1820, in the eyes of the mathematician and prefect Joseph Fourier (who had attended Laplace's Year III mathematics lessons at the *École normale*), the administration of registrations in France, the offices of the prefectures and the ministries, the know-how of clerks who were often incompetent but sometimes exceptionally zealous, the possible attention and rigour that Academicians of Sciences brought to this kind of calculation – all this formed a vast pyramid, in truth certainly imperfect but, at the very least, likely sooner or later to produce regular and sufficiently homogeneous accounts of population shifts. In common with his generation of scholars, and for the first time in history, Joseph Fourier was able to see as a coherent whole the administrative registration of births, the gathering of figures by a particular specialized office, the summarizing of tables by year or decade, the pinpointing of arithmetical regularities, and their use for other surveys or even their extrapolation in terms of a law of the phenomenon being counted. Although the new administration was quite imperfect, it was from then on subordinate to scholarly standards, so, in the eyes of a Laplace or a Fourier, it would be able to supply empirical materials that conformed to their expectations, allowing the science they were envisaging to be extended eventually to economic, political and moral issues.

Fourier and his colleagues intended to have a laboratory of moral sciences on a scale that would cover the whole of the new nation. It was to be the distinctive feature of a new social division of labour, producing the tools needed to carry on the activities of a state that would be independent of older, mainly religious, forms for legitimating political authority. From the early 19th century, this new division of labour allowed population counts, including birth records in particular, to be organized specifically and as systematically as was possible. Contemporaries therefore used the word "statistics", then being employed in France in a new way. The old political arithmetic was banished (Brian, 1994a).

This new configuration in the social division of administrative and academic labour opened up the immediate possibility of creating more rigorous and legitimate linkages between the registers compiled and mathematical calculations, without scholars having to agree amongst themselves on the subject in advance. On the contrary, for several decades, Laplace, Fourier and some others – that is, those scholars who were best informed as to the new possibilities and difficulties – explored them and confronted one another in a space all the more limited because it was characterized by a very rare competence… in other words, a very specialized scientific field.

From his Year III lessons at the *École normale* to the final revisions of his *Essai philosophique sur les probabilités* (Laplace, 1795, 1812, 1814) in the mid-1820s, Laplace continued to consolidate his analysis of the near-constancy of the ratio of the two sexes at birth, which he assessed at 22/21 [51.16%]. His major concern was the *general cause* of the phenomenon, in the same sense that D'Alembert had been able to explain the movement of the winds through the analytical calculus, taking into account only the most dominant physical phenomenon and relating local winds to particular causes (D'Alembert, 1747).[2] This time, the general cause was the

[2] D'Alembert's work, in response to a programme drawn up at the Berlin Academy, was done in the context of taking the calculation of the movements of fluids on the earth's surface a stage further than the works of the preceding decades on tides; it was characterized by the establishment of a hierarchy of causes. The hierarchy proposed by the Prussian Academy stated that "[t]he movement of the winds can perhaps be determined only by these three causes: namely, the movement of the Earth, the force of the Moon, and the activity of the Sun". Of course, D'Alembert resolved the issue through the skilful use of the analytical calculus and by stating the necessary hypotheses precisely from the point of view of celestial mechanics and the possible effect of heat from the Sun. But above all, he redefined the hierarchy of causes, subverting the order proposed in the initial programme and bringing his analysis of the hierarchy of causes into line with all his calculations (D'Alembert, 1747, pp. 8–9). "However inconstant the course of the winds may seem, it is nevertheless subject to certain laws. Navigators have long observed that the air has a steady movement out to sea in the torrid Zone; and if they have noticed some variations in this movement, this has been principally close to the coasts, and towards the places where the Ocean is narrowed by the Land. Therefore one cannot fail to recognize that among the different causes of the winds, there is at least one whose action follows a uniform, invariable order, and whose effects, even though they seem the most irregular, may not be modified and, so to speak, disguised by accidental causes. Thus the first object that a Philosopher must have in view when he sets out to go more deeply into the Theory of the winds, is to examine what may be this general cause and to determine by calculation, if possible, its quantity, its action and its effects." (D'Alembert, *ibid.*, pp. i–ii). Several decades later, this was to provide the model for Laplace's contribution. On the particular skills of geometers that combined the classificatory arrangement of the object and its reduction to the calculation, see Brian (1994a).

"greater facility in the birth of boys" (we would read in an anachronistic manner "the greater probability of a male birth") and its effect was the larger number of boys among births when considered in a large enough number.[3]

Laplace considered the measure of this expression – "in a large enough number" – to have been demonstrated in the papers he had written in his youth, and again in his *Théorie analytique* (1812). In fact, his calculations were particularly complex, and several more re-writings were needed, up to the 20th century – first for mathematicians to be able to talk about the "law of large numbers" (Poisson, 1835), then for them to agree precisely on the current terms, "strong law" and "weak law of large numbers", or "central limit theorem".[4]

This relates to the limits that are probable for the proportional variation of one case out of all possibles. It has long been common knowledge that there are gains in regularity with increases in observations, as witnessed by the expression "the strong carrying the weak" (Brian, 1991; Perrot, 1992). Jakob Bernoulli considered this when writing his *Ars Conjectandi* [*The art of conjecture*] (prepared between 1685 and 1705, published in 1713). For his part, Laplace proposed a demonstration to which he kept returning, and which provided the starting-point for the rapid development of the analytical calculus of probabilities (Hald, 1990; Stigler, 1986).

Each time he tackled the question, Laplace repeated his view that the excess of boys resulted from a constant law of nature, and that chance alone

[3] Laplace himself compared his theory of probabilities to the state of knowledge in the physics of fluids in his "important remark" at the end of the introduction to the *Théorie analytique*, a remark repeated in the conclusion to the *Essai philosophique* (1812, pp. clii–cliii of the 1886 edition; 1814, p. 105 of Vol. 2 of the 1921 edition; translated by Truscott & Emory from Laplace, 1951, p. 194).

[4] Laplace contented himself with this expression: "[...] that the ratios of the *effects* of nature are very nearly constant when these *effects* are considered in great number". He and his contemporaries, Fourier and Poisson, endeavoured to derive satisfactory proofs from this proposition, put forward in his *Théorie analytique des probabilités* (1812). Nowadays, mathematical formalism is different, very general and more powerful. Thus, a regularity (the one studied by Jakob Bernoulli, Laplace, Fourier, Poisson, Quetelet and others) is expressed by means of a series of results that must be placed in a hierarchy on a scale of increasing generality and in correlation with restrictive conditions: this refers to the Bienaymé-Chebychev inequality, to various laws of large numbers and to central limit theorems (Feller, 1950). It is important to clarify that mathematicians are still going more deeply into the convergence theorems in the theory of probabilities, and also that they frequently exercise themselves in attempting to give proofs of it that are as terse as possible (Kallenberg, 1997).

explained the variations known from observation. He invited scholars who were taking part in the 1798–1801 expedition to Egypt to check on this constancy; and he was delighted when Alexander von Humboldt's observations on South and Central America confirmed his position. Moreover, he noted that in Paris the proportion of boys was smaller, and saw this as the effect of girls born in the countryside being more frequently abandoned at the Paris Foundlings Hospital (see Box below).

Laplace's assessment (1795–1825)

One may draw from the preceding theorem* this consequence which ought to be regarded as a general law, namely, that the ratios of the *effects* of nature are very nearly constant when these *effects* are considered in great number. [...] I do not except from the above law *effects* due to moral causes. The ratio of annual births to the population, and that of marriages to births, show only small variations.

If we apply this theorem to the ratio of the births of boys to that of girls observed in the different countries of Europe, we find that this ratio, which is everywhere about equal to that of 22 to 21 [51.16%], indicates with an extreme probability a greater facility in the birth of boys. Considering further that it is the same at Naples and at *Petersburg*, we shall see that in this regard the influence of the climate *is imperceptible*.

At Paris the baptisms of children of both sexes deviate a little from the ratio of 22 to 21. Since 1745 [...] up to [...] 1784, there have been baptized in this capital 393386 boys and 377555 girls [51.03%]. The ratio of the two numbers is almost that of 25 to 24 [51.02%]; it appears then at Paris that a particular cause *brings* baptisms of the two sexes *close to* equality. If we apply to this matter the calculus of probabilities, we find that it is a bet of 238 to 1 in favour of the existence of this cause, which is sufficient to authorize the investigation. Upon reflection it has appeared to me that the difference observed holds to this, that the parents in the country and the provinces, finding some advantage in keeping the boys at home, have sent to the Hospital for Foundlings in Paris fewer of them relative to the number of girls according to the ratio of births of the two sexes. This is proved by the statement of the registers of this hospital. From the beginning of 1745 to the end of 1809 there were entered 163499 boys and 159405 girls [50.63%].

* Laplace is here referring to the proof of Jakob Bernoulli's results (1713) – later described by Denis Poisson as the "law of large numbers" (1835, 1837) – which he intended to establish using the generating functions.

Source: Laplace, *Essai philosophique* (1825 [1921]), extracts from Volume I, pp. 57–67; translation based on Truscott and Emory from Laplace, 1951, pp. 61–68: the italicized words indicate our revisions to this classic translation).

Laplace, throughout all his publications, made the proportion of the sexes at birth the topic to which he applied his new calculus of probabilities to the analysis of causes. But if we compare his pre-Revolutionary papers and his later publications, he appears to have gradually limited his understanding of the phenomenon to its most general aspect, which he made into a law whose sole elements are chance and a single cause: the greater facility of the birth of boys. This "law" focused on the excess of male births, and not on its variations. Its argument was above all mathematical in nature. The cases of Paris and the Foundlings were, in his eyes, merely minor physical disruptions, the second of which explained the first.

2. A SIMPLE RULE FOR PRACTICE

Two members of the expedition to Egypt, Joseph Fourier and Gilbert Chabrol de Volvic (1773–1843), both long-serving prefects (the first in Isère, the second in Paris), tackled the issue of the proportion of the sexes at birth during the early 1820s. They demonstrated some empirical scepticism towards Laplace's "law": they accorded more importance to the "physics" of the phenomenon, going as far as to organize its observation by the administration. At that time, the split was along the following lines: Laplace, like his disciple Denis Poisson (1781–1840), took the view that the highest priority for the issue was mathematical, and that its physical exploration was secondary. For Fourier and Chabrol, on the other hand, the only mathematical aspect was the matter of the rule to be adopted for calculation, and the highest priority had to be a focus on the development of empirical explorations. Moreover, these two positions matched their scientific orientations: Laplace embodied academic orthodoxy, to which Poisson was particularly faithful, while Fourier took a singular academic route, and his positions – although certainly recognized by posterity – were heterodox for their time. In addition, their opposition coincided with political divergences: the first two were satisfied with the restoration of the Bourbons to the throne, while Fourier remained a reformer born of the Revolution and the Empire. This configuration of conflicts was intensified by several other tensions, in the Academy of Sciences and at the *École polytechnique* as well as on other scientific terrains, like the theory of heat, for example (Dhombres & Robert, 1998; Armatte, 2004; Grattan-Guinness & Ravetz, 1972; Métivier, Costabel, & Dugac, 1981). It is important to observe that the tension between Fourier and Poisson (which was not a matter of public controversy) reflected not just two mathematical styles, two routes through and two assessments of the analytical legacy of the preceding generation, but also two possible pathways across the area of the mathematical and physical sciences where probability, precision and approximation intersected. Indeed, Fourier's physico-mathematics presupposes a homogeneity of phenomena that is not present in Laplace's or Poisson's work.

This explains Chabrol's and Fourier's involvement, during the 1820s, at the moment of publication of the *Recherches statistiques sur la ville de Paris [Statistical research on the city of Paris].* In the political context of the Restoration and the resulting relegation of administrative counting projects – a vast undertaking born of the Revolution and of the Empire – their intention was to "draw public attention to a large number of authentic facts of which it [was] important to spread knowledge" and, as an example, to publish a "general list of the questions that should be dealt with" (Chabrol de Volvic, 1821–1829).[5]

Fourier explicitly situated his four introductions to the *Recherches*, covering both results and methods, on a continuum with the endeavours of the reforming academicians at the end of the *Ancien régime*; while he cautiously confirmed Laplace's position on the regularity of the proportion of the sexes at birth, he also delivered a serious challenge to the general constancy of their ratio and supplied a technical rule for the guidance of observers (Fourier, 1821, 1823, 1826, 1829; see Box, p. 36).[6] In citing markedly different figures for London, Fourier was turning the young Laplace's results on the difference between Paris and London against the mature scholar.

[5] In the volume that appeared in 1823, Fourier explained: "This is how the administration has entirely fulfilled its plans, which were: firstly, to gather and to preserve all the earlier documents; secondly, to determine, through a general operation, the present state of the population; thirdly, to observe in an orderly fashion all the annual variations" (p. xvi).

[6] It is to be regretted that strictly mathematical historiography has not taken into account the two last papers to be included in Joseph Fourier's four introductions. Similarly, commentaries on the works of his rival Denis Poisson (see later) have been based on an equally restricted set of publications, to the detriment of others that the author had in fact mentioned explicitly. Such are the routines of a historiography of the sciences that accords more importance to decoding (often already well-known) texts retrospectively than to restoring them to their most literal forms (Perrot, 1992). We should add that Fourier's manuscripts are available, deposited in the *Bibliothèque nationale* (Fr. 22515, 22517 and 22518) (Charbonneau, 1993).

Fourier's empiricism (1821 and 1823)

In order to form a distinct idea of all the elements of the population, it has been considered in its permanent state; but it is evident that [its] constant law cannot be rigorously established. Population never subsists without some alteration, and it is very important to know the causes which modify it and those which would tend to preserve or re-establish it. First of all, it is recognized that population is subject to accidental and fortuitous variations which compensate for each other reciprocally over an interval of several years. [...] [The] indefinite repetition of events which are regarded as fortuitous makes all their variables disappear; in the series of an immense number of facts, nothing but constants and necessary ratios subsist, determined by the nature of things. [...] It is for a similar reason that the ratio of numbers of newborn infants of the two sexes remains more or less fixed in a given country. The physical causes whose involvement determines this result are entirely unknown to us: but the effect subsists, and it is free from uncertainty. [...] One of the least variable elements is the ratio of the number of births of boys to the number of births of girls. The first always exceeds the second, where both are fairly large, and their observed ratio in France differs little from that of 22 to 21 [51.16%] [...]. Observations seem to indicate that this ratio is not exactly the same in the various states of Europe. It is assessed at 19/18 [51.35%] for London, and at 21/20 [51.22%] for Naples.*

It will be decided if [the fact that the number of boys exceeds that of girls at birth] is a general one, suitable for all climates, when a large enough number of observations has been gathered in the diverse regions of the globe. [...] But what is the exact measure or the value of precision, and how does this quantity depend on the number of observations? This important question, which recurs in most research, belongs to the analysis of probabilities. It can be resolved through a fairly simple rule for practice, which we cannot explain here without deviating from the chief matter [...]. The precision of the mean result increases with the number of observations, and in proportion to the square root of this number.**

Source: Fourier, *Recherches statistiques*, *1821, pp. xxxvi–xli and **1823, pp. xviii et xxi-xxii.

From 1823, Fourier asserted that it was enough to take the view that the deviation around the proportion of births was proportional to the square root of the number of observations (which is the numerical form of what is nowadays called the central limit theorem). His two papers of 1826 and 1829 consisted of justifying, through mathematical proofs, what he viewed as a "fairly simple rule for practice", providing a degree of precision resulting from the accumulation of cases counted. In his mind, this was a very general rule specific to the numbers themselves, and as much applicable to the repeated measurement of the specific height of one pyramid at Giza as to that of a large number of comparable but distinct cases.[7] As we have seen, Condorcet before him had considered the repeated measurement of a certain fact and the proliferation of uncertain but comparable facts to be two similar questions (Condorcet, 1786).

Fourier was therefore opening a breach in the Laplacian monopoly over passing down the analytical calculus of probabilities from the late 18th century. He reactivated the empiricism of his predecessor, Condorcet, who had died prematurely during the Revolution. This led him, in one of his introductions to the *Recherches*, to wonder about the seasonal variability of the proportion of births according to sex, about the geographical origin of Parisian foundlings, and about how to establish the clear distinction needed between abandoned babies and illegitimate births, of which administrative figures for Paris revealed a predominant but smaller share to be male (Fourier, 1823).[8]

3. POST-REVOLUTIONARY AGRONOMY AND THE THEORY OF GENERATION

The Napoleonic period saw the appearance of a new group of scientific protagonists on the terrain of research relating to the proportion of the sexes at birth: agronomists. It is necessary to go back to the late 18th century, to a time when – as in Condorcet's work – a similar kind of thinking brought together speculation about the possibility of perfecting humankind, both physically and morally, and knowledge about the subject of improving

[7] On the "precision" of measuring a population phenomenon, see Fourier (1823, 1826, 1829); on Fourier's method and on the measurement of the pyramids at Giza, see Fourier (1829, pp. 569–570 of the 1890 edition) and Dhombres and Robert (1998, pp. 366–367). For a historical and epistemological approach to precision, see Wise (1994).

[8] Alain Bideau and Jean-Pierre Bardet, in Dupâquier (1988, pp. 373–398), highlight the fact that illegitimate births increased in the urban environment in France during the 18th century. The number of abandoned children also clearly increased, and even more significantly. So it is reasonable to take the view that the abandonment of legitimate children was a method of regulating the size of urban families. This increase (in the order of tenfold) was identified during the period in question, as is evidenced by summaries published in the early 19th century in ([Anonymous], 1808), which gives these figures: in 1640, 372 foundlings received at the Paris Hospital; in 1690, 1,509 children; in 1740, 3,150 children; in 1790, 5,842 children (pp. 42–43). On this subject, see also Weiner (1993). The distinction called for by Fourier therefore concerned a phenomenon known in his times, yet hardly touched on by Laplace.

domestic animal species. This was not a question of the theory of evolution of species nor of therapeutic or embryological concepts of procreation – two areas of thought that remained, for the time being, anachronistic.[9]

On the death of the great naturalist Georges-Louis Leclerc de Buffon (1707–1788), author of the *Histoire naturelle [Natural history]*, his academic eulogy gave the Permanent Secretary of the Academy of Sciences – Condorcet, who was hardly enamoured of his elder's style or method – the opportunity to mention the work of Abbot Lazzaro Spallanzani (1729–1799), who had recently become known in Paris for having succeeded in artificially inseminating a bitch using a syringe (Spallanzani, 1785). For Condorcet, the Abbot's experiments, like those of the Swiss medical doctor Albrecht von Haller (1708–1777), contradicted Buffon's general views. So much so that, some years later, Condorcet completed his famous prospectus, the *Esquisse d'un tableau historique des progrès de l'esprit humain* (1795), with the suggestion that asymptotic results were to be expected in the progress of humankind and, in particular, in empirical research in comparative physiology.

> "So, [...] we are bound to believe that the average length of human life will forever increase unless this is prevented by physical revolutions; we do not know what the limit is which it can never exceed. We cannot tell even whether the general laws of nature have determined such a limit or not. But are not our physical faculties and the strength, dexterity, and acuteness of our senses, to be numbered among the qualities whose perfection in the individual may be transmitted? Observation of the various breeds of domestic animals inclines us to believe that they are, and we can confirm this by direct observation of *humankind*." (Condorcet, 1795; p. 458 of the 2004 edition; based on Barraclough's 1955 translation, from Baker, 1976, p. 280: the italicized word indicates our revision to this classic translation)

This passage heralded developments in his arguments on the future of humanity, which were published only in 1922: "Fragment 10", which we have already discussed in Chapter 1 above (see also Appendix A). But Condorcet posthumously gave the public other indications in the 1804 edition of the *Esquisse*. These were primarily in *Fragment 9* ("*Sur l'Atlantide*

[9] Charles Darwin, as is well-known, published his two main works during the second half of the 19th century. On the medical aspects of artificial reproduction, see Delaporte & Delaporte (2004).

[On Atlantis]"): a fresco of a scholarly world of future times, organized collectively in the hope of combating contemporary prejudices:

> "Finally, there are endeavours to which, either by the very nature of the object or by its smallness or by the extreme uncertainty of success, one man alone may fear to devote himself, because he would expose himself to ridicule or to a kind of shame: and even when one feels that one could receive the imprint of such brands only from the hand of a contemptible prejudice, who knows how much they are still feared, how much dread is felt of the opinion of those very people whose reason is most despised? [...]. I shall place in this class the research begun by Spallanzani on generation that might be described as artificial, or research into the causes that determine sex, either in the foetuses of viviparous animals, or in the germs of eggs." (Condorcet, 1804; p. 910 of the 2004 edition)

On stud-farms, for example, the formal recommendation was for multiple cross-breeding, with a view to obtaining the best results. Around 1800, the most advanced practice, all over Europe and notably in England, France and Germany, consisted of attempting to control the movement of breeding animals and to organize the registration of coverings and of the characteristics of stallions, mares, and their offspring (Lafosse, 1775). In the racing world, in England from the late 18th century onwards and in France from around 1830 (Bryon, 1828; Weatherby, 1791), in addition to these practices, there was registration in the form of Jockey Club stud-books, with the genealogy of stallions and of runners, as much in order to avoid fraud as to meet the requirements for betting (Blomac, 1991).

In France, agronomists and physiologists conducted more in-depth research on the generation of domestic animals, notably that of sheep, cattle and horses. This resurgence of interest, supported by the government under the First Empire, should be viewed in the context of early-19th-century wars in Europe, which consumed an inordinate number of horses for cavalry, service corps and staff, and vast quantities of wool and leather for clothing and bedding, and finally of meat for provisioning. However, it still remains for historians to measure the effects of these continent-wide upheavals on regional varieties of domestic species. It is thus reasonable to consider that most of the "local races" encouraged in the 19th century were pure inventions resulting from this continental military and economic cataclysm.

Among the scholars active in this sphere can be included two seasoned experimenters, one a monarchist established not far from Paris, Charles-Gilbert Terray de Morel-Vindé (1759–1842),[10] the other a Bonapartist settled in Aveyron, Charles Girou de Buzareingues (1773–1856).[11] Particularly involved in the section of the First Class of the Institute called "Rural Economy and Veterinary Medicine",[12] they met in the circle of the veterinary scientist, Academician Jean-Baptiste Huzard (1755–1838), where their work on merino sheep placed them to some degree in competition. Both authors had the same target: the knowledge of country stockbreeders, whose maxims they assessed through systematic observations and commentaries. At the Academy of Sciences – where both were corresponding members and one was later on a full member – they found a home to register their reports and a place where their agronomy could become an established science.

Morel-Vindé published, in several instalments from 1813 to 1816, his *Observations sur la monte et l'agnelage [Observations on breeding and lambing]* in the *Annales d'agriculture française*. For each year, he gave a detailed register of births on his sheep farm and the characteristics of his breeding animals (Morel-Vindé, 1813). In the aim of undermining commonly-held beliefs about the influence of the age of ewes on the sex of their descendants, he supplied tables on the distribution of births for the years 1812 and 1813, distinguishing year by year, the age of the mothers and the two sexes of the offspring. The distinctive feature of his study, carried out using the same framework over several years, was to organize the numbering of each animal and make a collection of historical reports associated with each number. An instigator of the breeding of merino sheep, Morel-Vindé decided to take stud-farms as a model. This was because, at

[10] A magistrate during the later days of the *Ancien régime*, Morel-Vindé withdrew from public life during the Revolution in order to devote himself to agronomy, while always remaining a constitutional monarchist. He was a corresponding member of the Rural Economy Section of the First Class of the Institute from 1808, then one of its members within the 1824 Academy of Sciences, as well as a member of the Agricultural Society. He was made a peer of France in 1815, and joined the Higher Council of Agriculture in 1818. He carried on his breeding practices at his La Celle Saint-Cloud property (Girardin, 1845).

[11] A Bonapartist and supporter of the Empire even during the Hundred Days in 1815, Girou de Buzareingues was a corresponding member of the Rural Economy Section of the First Class of the Institute in 1826 and a member of the Agricultural Society. In December 1828, he failed in his attempt to succeed to Louis Bosc d'Antic's chair at the Academy of Sciences, encountering opposition from the Rural Economy Section (see Brian & Jaisson, 2005d).

[12] The First Class of the Institute of France derived in large part from the old Royal Academy of Sciences, and reverted to that name from 1816.

the time when he published his work, only human beings and horses were the object of individual birth registers, set up by churches or civil register offices for the first, and by breeders and the authors of stud-books for the second. The advantage of the method was that annual statements were obtained, individual by individual, and tables were drawn from them that enabled conclusions to be made on a variety of conjectures. In the 1814 article, and for the first time, a list of ages was drawn up for experimental purposes from an *ad hoc* register of the mothers; starting at eighteen months old and continuing from year to year, it was then cross-tabulated with a distribution by the sex of births.

It was in a completely different journal, the *Annales des sciences naturelles*, that Charles Girou de Buzareingues published "*Observations sur les rapports de la mère et du père avec les produits, relativement au sexe et à la ressemblance [Observations on the relationships of the mother and the father with their products, relative to sex and to resemblance]*" (1825). He reviewed Morel-Vindé's compilations in passing, in order to draw from them three annual tables of the sex distribution of lambs according to the age of the ewes. On this occasion, Girou de Buzareingues proposed a distinction between three cases: that of "middle"-aged ewes (nowadays, these would be described, more precisely, as being of modal age), and those of older or younger mothers. From there, Girou de Buzareingues gave recorded figures based on his predecessor's, both supporting and contradicting them, showing the age of the mother as having an effect on the sex of her descendants. We should point out that Girou paid no heed to the small total number in the median category, nor to the fact that the two sexes, in general, were not necessarily equal.

Annales des sciences naturelles was a new journal, created a year earlier and led by a dynamic editorial team consisting of three brothers-in-law who were the heirs to a dynasty of scholars active since the late 18th century and especially under the First Empire: Jean Audouin (1797–1841), Adolphe Brongniart (1801–1876) and Jean-Baptiste Dumas (1800–1884), who were all aged between 24 and 28 in 1825.[13] As editors, this trio of young turks

[13] Jean Audouin, an entomologist, was to be elected as a member of the Academy of Sciences in 1838; Adolphe Brongniart, a botanist, would become an Academician of Sciences in 1834; Jean-Baptiste Dumas, a chemist, was to be a minister under the Second Empire; elected as a member in 1832, he would become Permanent Secretary of the Academy in 1868. They were the son and sons-in-law of Alexandre Brongniart (1770–1847), mining engineer and zoologist, elected as a member in 1815, himself the son of the architect Alexandre-Théodore Brongniart (1739–1813) and the son-in-law of

used the collective signature "R." (for *Rédaction*: "Ed.") to nine pages entitled "*Speculative summary*" ("*Résumé spéculatif* ", 1825) immediately before Girou de Buzareingues' seventeen pages of observations; the text also referred to the mathematician Joseph Fourier's recent statistical research, which had appeared in *Recherches statistiques sur la ville de Paris* (1821, 1823). At the same time, the paper was punctuated by a short *nota bene* giving a renewed justification for the publication of the paper itself and of the "speculative summary". The whole was further extended by an extract from a report that had already appeared – this time, one that challenged some of the conclusions attributed to Fourier.[14] The combination of these means aimed to make it possible to publish the "speculative summary". We can guess that the young editors were struck by the potential of the combinatorial analysis that they had extrapolated from Girou de Buzareingues' observations. It is true that their commentary detailed what nowadays and in retrospect may appear to be a prototype for the cross-tabulated distribution of a nominal variable according to two analogous centred variables. Seen with the mathematical eye of a 20th-century statistician, one would tend to describe it as the beginnings of "experimental design" (Fisher, 1925; cf. Hald, 1998). However, there are no test measurement issues: it was only a matter – although a considerable matter – of highlighting variability in generation according to a combination of the same two measures applied to the father and to the mother. Turning now to the point of view of the history of biometry, we know that these different elements were to be present sixty years later, in 1886, when Francis Galton (1822–1911) perfected linear regression – still in relation to the question of the transmission of physical qualities in the process of human generation (Hald, 1998; Stigler, 1986).

Girou de Buzareingues thought he observed that "middle"-aged ewes produced lambs with a balance between the two sexes, and that older or younger ewes produced more males. "R." generalized the records according to the following schematic, simplified formula: old mother, frequent male sex; middle-aged mother, equal ratios; young mother, frequent male sex. They extrapolated from this the additional formula: old father, frequent female sex; middle-aged father, equal ratios; young father, frequent female sex. Then they combined the two according to the principle of adding the parental influences together. We can reproduce the list of the nine possible

---

Charles-Etienne Coquebert de Montbret (1755–1831), mining engineer, elected to the Academy in 1816.

[14] This was Bailly (1825). The extract comes from the *Bulletin de la Société de Philomathique*, October 1824; the Brongniart family was active in this society.

cases they envisaged, in a manner that is more familiar nowadays – i.e. cross-tabulated (see below).

	Young mother	Median-aged mother	Old mother
Young father	Equal ratios	Frequent female sex	Equal ratios
Median-aged father	Frequent male sex	Equal ratios	Frequent male sex
Old father	Equal ratios	Frequent female sex	Equal ratios

Similarly, "R." generalized the other criterion discussed by Girou de Buzareingues: the parents' state of health, applying the doxic principle that expects strength to beget strength – strong mother, frequent female sex; medium mother, equal ratios; weak mother, frequent male sex; etc. This gives a new combination, which takes the following form.

	Strong mother	Medium mother	Weak mother
Strong father	Equal ratios	Frequent male sex	Frequent male sex
Medium father	Frequent female sex	Equal ratios	Frequent male sex
Weak father	Frequent female sex	Frequent female sex	Equal ratios

After the 1825 publication of the article, with its recklessly broad assertions, Girou de Buzareingues continued to accumulate empirical observations and general views, and to put them forward to thc Academy of Sciences, of which he was a corresponding member. So it was that he read the opening sections of a *Mémoire sur la génération [Paper on generation]* to the session of 17 January 1825. According to the minutes of the meeting, the mathematicians Laplace, Poisson and Fourier attended this lecture, and the veterinary scientist Jean-Baptiste Huzard was appointed to assess the paper.[15] Two years later, early in 1827, Girou sent eleven new dispatches to this Company. The series did not end there; and it was also punctuated by lectures read by Girou himself or on his behalf. His tenacity reached its

[15] Institut de France, Académie des sciences, *Procès verbaux*, vol. VIII, p. 483.

culmination when, on 10 November 1828, he offered the Academicians a *Mémoire sur la distribution et les rapports des deux sexes dans le Royaume [Paper on the distribution and ratios of the two sexes in the Kingdom]*, in which he analysed French register offices figures for the preceding decade. He concluded on the principle that: "the male sex results from the predominance of the driving force" (Girou de Buzareingues, 1828, p. 309). He summarized his earlier writings in a book entitled *De la Génération [On generation]*, and, a month later, on 15 December 1828, he submitted it for the Academy's prize founded by Montyon, to be awarded to a work of experimental physiology; at the same time, he put himself forward for election to the place left vacant after the death of the agronomist and naturalist Louis Bosc d'Antic (1759–1828). Girou de Buzareingues was not elected, and the prize committee's report, read on 8 June 1829, awarded him an honourable mention for his experiments – in last place, after the prize and five other mentions.[16]

4. THE BLADE FALLS IN PARIS: POISSON'S CALCULUS OF PROBABILITIES

In the meantime, others had taken up the torch of the Marquis de Laplace, even before his death on 5 March 1827. The publication of *Recherches statistiques sur la ville de Paris*, edited by Joseph Fourier, and most particularly of its second volume, dated 1823, elicited a reaction from those close to the aged Laplace, led by Denis Poisson. A veritable patchwork of articles appeared, primarily in the periodicals of the Office of Longitudes (its *Connaissance des Tems* and its *Annuaire*), where the Laplacians could escape the control of the Academy's Permanent Secretariat – in which, from late 1822, Joseph Fourier himself was Secretary for the division of Mathematical Sciences (Poisson, 1824; Poisson & Nicollet, 1824 et seq.).

Some weeks after the culmination of Girou de Buzareingues' insistent approaches, Poisson submitted a paper to the Academy of Sciences, in which he reformulated Laplace's analysis of the error dispersion for distribution of the sexes at birth. On 9 February 1829, Poisson gave a reading of his paper *Mémoire sur la proportion des naissances des garçons et des filles [Paper on the proportion of births of boys and of girls]* (Poisson, 1830)

[16] These events have been traced, as have those which were to be commented on later both in the minutes of the Academy of Sciences and in the archives of this learned society. See Brian & Jaisson, 2005d.

before an assembly of the Academy. As usual, no fewer than five or six members of the Mathematics Section and a similar number from Astronomy were present, and, according to the minutes, a dozen of those in attendance had connections with the Office of Longitudes; however, by some unlikely chance, the Permanent Secretary, Joseph Fourier, was absent.

Poisson relied on the regular publication by the Office of Longitudes of figures for civil registrations of births according to sex and legitimacy. He continued by giving a long proof that has remained famous in the history of the analytical calculus of probabilities, as a great moment in its mathematical development (Hald, 1998, pp. 230–242 and pp. 571–575).

> "But in order for the formulae of the calculus of probabilities in question to be independent of the law of probability of deviations that has not been given to us, observations have to have been made in considerable number; this does not allow these formulae to be applied to research on the ratio of the annual births of both sexes, of which we know well only the ten values observed in France from 1817 to 1826." (Poisson, 1830, p. 308)

This was the opportunity to clarify the question of estimating the parameter of a binomial law. Poisson then asserted that observations of the registers over a ten-year period had led to a precise estimate. He indicated how to estimate, through the calculus, the uncertainty to which it must be subject.

On 4 May 1829, probably as a response to Poisson, and before the annual announcement of prizes, Girou de Buzareingues returned to the charge, submitting to the Company's judgement a new *Mémoire sur la distribution des mariages, des naissances et des sexes dans les divers mois [Paper on the distribution of marriages, births and the sexes in diverse months].* He announced that his experiments on generation had been confirmed, hoping to see this confirmation subjected to examination by a new committee of the Academy. The mathematician Joseph Fourier and the physiologist François Magendie (1783–1855) were appointed for this new evaluation. But the former died on 16 May 1830; more than a year later, on 1 August 1831, Girou presented his regards to the Academicians and asked for a new committee to be appointed. Finally, on 19 September 1831, a committee made up of the engineer Pierre-Simon Girard (1765–1836) and the astronomers Charles Damoiseau de Montfort (1768–1846) and Louis Mathieu (1783–1875) read a highly descriptive report to a session of the Academy. They concluded by inviting the agronomist "to continue his interesting research". On 22 August, a month earlier, Girou de Buzareingues had read his paper *Mémoire sur le rapport des sexes dans le règne végetal [Paper on the ratio of the sexes in the plant kingdom].* Yet, again on 19 September, another report from the Academy also demonstrated a similar reserve towards this work. Girou de Buzareingues' account with the Academy seems to have been closed on that day. However, the agronomist remained a zealous corresponding member of the Academy until his death. Statisticians and mathematicians were to return to these questions again only along the biometric route opened up by Galton more than fifty years later. In the meantime, the consolidation of the theory of the average by Adolphe Quetelet (1796–1874) would be necessary before – once the measure of central tendency had been consolidated – concomitant variations of dispersions around this trend would lend themselves to subsequent statistical calculation.

Let us return to Poisson's paper published in 1830. With this and two other works (Poisson, 1835, 1837), the mathematician rescued the analytical theory of probabilities from the singularity of its first Laplacian expression. They led Poisson to what he called the "law of large numbers", formulating a proof which has of course required various later revisions up to the present day (Hald, 1990, 1998; Stigler, 1986, 1999), but establishing a clearer distinction than in Laplace's and Fourier's work between the probability of one sex at birth (what Laplace had called "the greater facility") and that of the measurement of a proportion of one sex in the total number of observed cases (the probability of observing the result under study). This is because two very different probabilities are in play in the matter: one on the side of cause, the other of effect.[17] From his 1824 and 1829 papers onwards, Poisson showed why, from a strictly mathematical point of view, deviations in the proportions observed should be measured with an order of magnitude that is the inverse of the square root of the number of observations. This was, therefore, a mathematical response to the argument of Fourier's "rule for practice".

Poisson worked on the figures for registrations of births gathered by town halls and at the level of *départements*. The *Annuaire* published by the Office of Longitudes was an official document and aimed for nationwide

[17] Using more recent terminology, the sex of each newborn baby is viewed as a binomial variable that takes the value "boy" or the value "girl"; and the proportion of births observed for a population under consideration is a real continuous random variable that is between 0 and 1 (for example, it is expressed through a percentage). The second variable is an estimator of the first. Poisson's proof established a relationship between the probable deviation of this estimator and the number of observed cases (Poisson, 1824, 1830). On this subject, see Appendix D.

dissemination of weights and measures standards and information about astronomical and natural phenomena, which could be used to regulate everyday life. Starting from the volume for 1825, it included a two-page notice entitled *Observations relatives au nombre de naissances des deux sexes [Observations relating to the number of births of the two sexes].* This repeated the same conclusions year after year, simply updated with the last known figures. The proportion of boys was in the order of 16/15. The variations between *départements* in the South of France and the North could no longer leave any doubt that there was no climatic effect. It was merely noted that natural births (that is, those outside marriages contracted under the Napoleonic Civil Code) showed a significantly lower proportion of boys than births as a whole.

This fact, likely to baffle scholars (Poisson, 1830), became immediately blurred in the eyes of the public by a handful of others produced one after another: "in these same thirteen years, there were nineteen occasions on which annual births of girls exceeded those of boys in some *départements*". There followed a list of these "less virile" *départements*, some of which were even habitual offenders. This was exactly the type of conclusion that Poisson told the Academy of Sciences was irrelevant, even though the *Annuaire* did not make that clear. From all this, the attentive citizen or the zealous civil servant could only conclude that scholars, as soon as they had enough figures, regarded the proportion of the sexes as constant... and that, as a citizen of his *département*, he would have to maintain both his and its standing. Morality was intact, despite the anomaly of natural children (Poisson & Nicollet, 1824 et seq.).

In his papers intended for a scholarly audience, the mathematician was more rigorous (Poisson, 1824, 1830). He rescued the principle of Laplace's calculuses and rejected conclusions based on total numbers that were too small. Nevertheless, he also had to concede that the ratio of births held by Laplace and by Fourier to be true enough... was false: it should be confined to 16/15 (51.61%) and not 22/21 (51.16%). He also recorded the fact that two particular cases stood out from the generally consistent picture, and the modern calculation derived from his own method confirms this (see Table 1).

Births in Paris and natural births for the whole of France both showed proportions of boys that were certainly larger than those of girls, but rather lower than those for the kingdom overall. This justified the identification

Table 1: Poisson's figures (about 1830)

	France: Total	France: Legitimate births	France: Births outside marriage	Paris: Legitimate births	Paris: Births outside marriage
	1817	1817	1817	1815	1815
	1829	1829	1829	1827	1827
Total (N)	12 577 421	11 693 486	883 935	215 639	122 404
Boys (M)	6 484 656	6 032 787	451 869	109 973	62 239
	(Poisson & Nicollet, 1832)			(Poisson, 1830)	

Results of the modern calculation based on the normal approximation to binomial distributions, drawn from Poisson's work (see Appendix D)

M/N	51.56%	51.59%	51.12%	51.00%	50.85%
$2/\sqrt{N}$	0.06%	0.06%	0.21%	±0.43%	±0.57%
Probability	95.45%	95.45%	95.45%	95.45%	95.45%
Maximum	51.61%	51.65%	51.33%	51.43%	51.42%
Minimum	51.50%	51.53%	50.91%	50.57%	50.28%

M/N	51.59%	On the left are the results of calculations made by Poisson himself, for the years 1817 to 1826, that is $N = 9,656,135$ and $M = 4,981,566$ (Poisson, 1830). Although the deviation in the confidence interval was reasonable, the calculated probability was manifestly wrong.
Limits	±0.07%	
Probability	99.9978%	
Maximum	51.66%	
Minimum	51.52%	

of two particular cause (see Box, p. 49). Moreover, Poisson again differentiated himself from Laplace and from Fourier by highlighting the fact that illegitimate births could not explain the situation in the capital.

Therefore, reading these texts, which had their origin in the Academy's desire to brush aside the issue of animal generation, gives us a record of a phenomenon that is one of the most general (the probability of one of the two sexes at birth is like a game of chance at a given level of probability – that is, the binomial model), of a technique for putting deviations in the proportions of the sexes at birth to the proof (the process of evaluation using approximation to binomial distribution), and of two different cases established as distinct from the general case (that of Paris, and that of births outside marriage).

Poisson's clarification (February 1829 [1830])

In considering the births of the two sexes over six consecutive years [...], I noticed four years ago: 1st, that the ratio of births of boys and girls was 16/15 [51.61%], instead of 22/21 [51.16%], as was previously believed; 2nd, that this ratio was well-nigh the same for the South of France and for France as a whole, so that it seems to be independent of variation in climate, at least across the whole country; 3rd, that its value, among children born outside marriage, was significantly less than for legitimate children, and almost equal to 21/20 [51.21%]. [...]

Nor is the proportion of births of the two sexes the same in Paris and in the *départements* [...]. It may be [...] presumed that in a great capital like Paris there also exists a particular cause that diminishes the preponderance of male births and that acts on both legitimate and natural children.

Our minds are naturally led to accept the results of the experiment with all the more confidence because they have been deduced from a greater number of observations; but if we want to assess probability and to know that of their future reproduction, we are obliged to use the formulae that mathematical analysis provides for this object: the perfection of these methods in general and their application to the facts I have just cited are the matter of the Paper that I am presenting to the Academy today.

We may [...] conclude that, in the present period [1817–1826] and for the whole of France, the probability of a male birth is affected by only very small variations from one year to another, and we may take for its value, the mean of the ten years that we have considered, that is, 0.5159 [51.59%]. As we are in ignorance as to the cause that makes births of boys preponderant, experience alone will be able to decide whether this probability will subsequently vary more, or remain almost constant. Observation has not yet taught us whether it changes with the seasons in the same year; nor do we know whether it is the same in different nations; we know only that it depends on the state of society, since the number of births outside marriage significantly influences the proportion of male and female births.

Source: Denis Poisson, *Mémoire sur la proportion des naissances*, (1830), pp. 239, 242 and 307.

Historians of mathematics have taken no notice of these statements, and historians of the social sciences have not even seen them... Yet it must be noted that Poisson was very clear: one of these particular causes, the illegitimacy of births, is known only through their probability and "we know only that [this probability] depends on the state of society, since the number of births outside marriage significantly influences the proportion of male and female births" (Poisson, 1830, p. 307). In opposition to Fourier – but not Girou – his radically mathematical point of view meant that Poisson was wary of the quest for a "moral physics". He took it in the direction of a dazzling yet very simple formulation: the elementary phenomenon – the sex of a birth under consideration – is adequately described by a general probability; the latter depends on the state of society; its manifestations are necessarily probable and measured as such.

From the retrospective point of view that we are introducing here, we arrive at an observation which would probably astonish our protagonists. In the manuscript of *Fragment 10*, which remained unpublished until 1922, Condorcet envisaged the possibility of humankind acting on the level of probability of births of the two sexes. Laplace's writings express nothing that would relate to such a concept. Everything points to Fourier wanting to continue certain aspects of Condorcet's work, even though he was apparently unaware of the *Fragment* on births. Poisson, on the other hand, formulated this somewhat radical idea – let us call it a probabilistic social phenomenon – not by relying on Condorcet's work itself, but by countering the physicalism of the person (Fourier, that is) who was inspired by it in defence of the scientific legacy of Laplace – who had himself been Condorcet's former rival, yet had been forgetful of him...

5. ENGLISH AND GERMAN WAYS

It is known that Condorcet's *Esquisse d'un tableau historique des progrès de l'esprit humain* (1795) was one of the main targets of the Reverend Thomas Robert Malthus (1766–1834), on the first publication of his *Essay on the Principle of Population* in 1798. Having read the *Esquisse* – which was a prospectus for the *Tableau historique* – the Anglican priest had rejected the idea that humanity could expect an improvement in its fate as result of population growth and an increase in the length of human life. This was because, for Malthus, population growth took place in geometric progression, while growth in resources was in arithmetic progression. This dual logic led him to anticipate a catastrophe which could be avoided only by limiting population growth, with differentiations according to social class. Malthus, in 1798 – that is, when the spirits of the English public were at their lowest ebb, during the war with Revolutionary France – wanted to stigmatize the trust in the progress of the human mind argued by Condorcet, to proclaim anew the Revelation and, in consequence, to assert that "excitements from intellectual wants [are] continually kept up by the infinite variety of nature, and the obscurity that involves metaphysical subjects" (Malthus, 1798, Chapter 19). The mysticism set in motion by the Reverend's *Essay* was thus a response to the deep thinking on the asymptotic perfectibility of humankind and its mind that had run throughout Condorcet's *Esquisse* some years earlier. Both the Fragments of the *Tableau historique* and the *Esquisse* suggest that Condorcet, if he had observed it, would have recognized "the hand of a contemptible prejudice" in such obscurantist reaction (see extracts from *Fragment 10* in Appendix A).

The Reverend's bleakness has been criticized by English post-Malthusian authors. Among them, a generation later, can be included Michael Thomas Sadler (1780–1835), a Tory parliamentarian: meaning that he was conservative, anti-liberal, attached to a religiously-inspired concept of politics – and therefore unlikely to refer back to Condorcet. An activist in the Poor Law movement in the social Christian vein, he published *The Law of Population* (Sadler, 1830). In opposition to conjecture about "super-fecundity", his work combined two ideas about population that had been distinct until then: one following both versions of Malthus' essay (1798 and 1803) but countering his pessimism, and the other born of reading Süssmilch (1741). For a providentialist, the differential between the sexes at birth was caused by divine anticipation of losses of males, in order to arrive at the right number for monogamy. Hence, Sadler had one chapter entitled "Of the law of population: anticipatory computations of Nature, especially in reference to the proportion of the sexes" (Sadler, 1830, pp. 332–351).[18]

Sadler, who had access to the family registers of a population of nearly a thousand peers of the realm, scrutinized the proportions of births of the two sexes according to the absolute ages of the parents, their relative ages, and – a rather unusual experimental device designed to focus on providential Nature's capacity for anticipation – cases of remarriage after being widowed. When Quetelet later made use of the table derived from this scenario, he was very careful not to indicate to his reader the political and theological

[18] We should note that, in 1829, the agronomist Morel-Vindé, mentioned above, published a piece of anti-Malthusian writing – more precisely, a critique of the theses of the French political economists who, during the 1820s, proclaimed the doctrine of Malthus. In it, he displayed himself as a defender of the French *laisser-faire* law ensuring freedom of territorial ownership, through his conviction that Malthus' conclusions were suited only to turn-of-the-century England and Ireland, not to Restoration France, whose population had incurred sufficiently heavy losses in the Napoleonic Wars to be spared a Malthusian-inspired policy. This work was neither explicitly nor implicitly within the scope of the English debate taking place in those years – the one in which Sadler was involved.

intent of the Tory reformer who, sustained by the Industrial Revolution and motivated by Parliamentary Poor Law debates, meant to safeguard the action of God, which was in Sadler's view wholly favourable. In concluding his chapter on the calculations that followed, he wrote: "We have abundant proofs as to the benevolent purposes of the Deity." (Sadler, 1830, p. 351)

Now we should go back across the Channel and then on over the Rhine, to find yet another system of relations between the same sciences, still with the same ingredients, producing results that were very little different yet more or less free of any academic tensions between mathematics and empirical counting. The first instance was a thesis in medicine, entitled *De qualitatibus parentum in sobolem transeuntibus, praesertim ratione rei equariae [Concerning the qualities transmitted by parents to offspring, especially with regard to equariae]* (Notter, 1827), defended at the University of Tübingen on 1 September 1827, by Friedrich Notter (1801–1884), who is not known today for any scientific works, but for his contribution to literature as translator of Dante and Cervantes into German, as a journalist and as a liberal politician of the 1848 period.[19] Next came a publication in German, dated 1828, whose title was almost a translation of the above: *Ueber die Eigenschaften welche sich bei Menschen und Thieren von den Eltern auf die Nachkommen vererben, mit besonderer Rücksicht auf die Pferdezucht.* This time, its author – according to a number of commentators and a good many libraries – was said to be physiologist and professor of veterinary medicine at the same University of Tübingen, Johann Daniel Hofacker (1788–1828). The latter had in fact presided over – nowadays, we would rather say he had supervised – the thesis defended some months earlier by Friedrich Notter. The new doctor appeared in the German version as a collaborator in the work; the body of the text of the two versions, Latin and German, is almost identical. Curiously, Hofacker's preface to the second is dated 10 September 1827: in other words, the publication in German – for a broad audience and with an explicit reference in the title to the significance of its conclusions for humankind – had been decided upon, at the latest, in the days following the initial viva examination. Was this a matter of plagiarism? Applying this criterion seems anachronistic: usage then required that those who were to receive a qualification had to defend their mentor's theses before the university.

[19] *Neue deutsche Biographie*, Berlin: Duncker and Humblot, 1999, vol. 19, pp. 366–367. Despite the singular path taken by this work, there is no doubt as to the identity of its author.

Be that as it may, it must be observed that Johann Daniel Hofacker was paying a great deal of attention to what was then happening in Paris. Thus, in 1826, he was responsible for the German translation of François Magendie's *Précis élémentaire de physiologie [Elementary handbook of physiology]*. He – and the students over whose theses he presided – studied the anatomy and physiology of domestic animals and of the human species. This collective programme of work, apparently interrupted by the premature death of its leader in 1828, was an astonishing echo, even in the title of Notter's thesis, of the final paragraphs of Condorcet's Tenth Epoch in the *Esquisse*. In fact, it was in Tübingen itself that the first German-language edition of that work had appeared (actually its first foreign edition), in the now fairly well-known context of the interest taken by young Swabian university academics in the French Enlightenment and the French Revolution.[20] The two versions of Notter's thesis relied firmly on Buffon, and explicitly cited Volume 22 of his *Histoire naturelle*, which in fact opens with Condorcet's *Éloge de Buffon [Eulogy to Buffon]* (1791), where, as early as 1788, he had revealed the programme later laid out in the *Entwurf* (German translation of the *Esquisse*), which appeared in 1796.

Notter's thesis carries the necessary tribute to the academic authorities, as well as a quotation from Shakespeare; like the German version, it takes as its starting-point (for Chapter III) Girou de Buzareingues' article and the "speculative summary", which appeared in 1825. It mentions in passing that Morel-Vindé had been the first to put the criterion of maternal age to the proof. However, Notter did not pay any particular attention to distinguishing Girou de Buzareingues's initial paper from the speculations of the editors of *Annales des sciences naturelles*; if so he attributed to Girou the combinations that the young commentators had introduced. However, Notter improved on the latter by drawing a true experimental design from their interpretation and applying it – to the letter – to the study of family registers in the town of Tübingen and a neighbouring village. For Tübingen, Notter considered 21 cases: the first six corresponded to six configurations of relative parental age gap; the next three singled out young fathers, those of median age and elderly fathers; three analogous headings were then applied to mothers; nine others supplied cross-tabulations of these two classifications.[21] The survey concentrated on

[20] Condorcet, *Entwurf eines historischen Gemähldes der Forschritt des menschlichen Geistes*, Tübingen, Cotta, 1796 (translated by Ernst Ludwig Posselt).

[21] Case No 1: the mother is older than the father; Case No 2: both parents are the same age; Case No 3: the father is from one to three years older than the mother; Case No 4: the father is from four* to six years older than the mother; Case No 5: the father is from seven* to nine years older than the mother; Case No 6: the father is from ten* to twelve years older

about 2,000 cases – a laudable effort – and the results were presented with caution. Even so, they fell foul of the mathematical critique that Poisson was to formulate elsewhere in 1829 and publish in 1830.

Hofacker did not content himself with proposing to Tübingen's principal academic publisher that a translation of this innovative thesis into German be published under his own name. Even from before the viva examination, he had wanted to alert scholarly Germany to his authorship: he therefore wrote to Johann Nepomuck Ehrhart von Ehrhartstein, editor of the *Medicinisch-chirurgische Zeitung*, published in Innsbruck. Thus, in the very last pages of the supplement to the final issue of 1827, Ehrhart published an extract from his Tübingen colleague's communication. In contradiction to the passages of Notter's thesis where Girou de Buzareingues' results had been compared case by case to those obtained from the Tübingen registers, Hofacker seems to be claiming the discovery as his own.

> "At the same time, I am seizing this opportunity to offer you a few notes on a dissertation shortly to be published under my supervision, in the hope that they might be suitable for an announcement in your journal. In this paper, which deals with the qualities that parents transmit to children, I have naturally covered the determination of sex by different momenta. For various reasons, I have already long thought that age was an important momentum in that regard. I have extracted from the registers of baptisms for 2,000 infants the indications relating to their sex and to the ages of the father and of the mother, and I found the following principal results [there follow 15 cases with figures].[22] All these observations and calculations were established with the greatest precision [...]. Manifestly, here are completely new laws, towards whose discovery I was led by chance and by some perseverance in calculation." (Hofacker, 1827, pp. 398–399).

---

than the mother; Case No 7: fathers aged 24 to 35*; Case No 8: fathers aged 36 to 47*; Case No 9: fathers aged 48 and over; Case No 10: mothers aged 16 to 25*; Case No 11: mothers aged 26 to 35*; Case No 12: mothers aged 36 and over. The nine other cases, numbered 13 to 21, are a combination of the six preceding ones, Nos 7 to 12 (Notter, 1827, 1828, Chapter III; for this transcription we have corrected the interval boundaries – marking them with an asterisk* – so that they do not overlap, according to the list of cases; in fact, most scholars of the period did not pay attention to that aspect of their tables, but made unambiguous calculations). To consult Notter's numerical results, see Brian & Jaisson (2005d, pp. 96–97). Apart from supplementary results drawn from the Hagelloch family registers, the thesis also presents data from perusals of the Marbach Stud books. However, these compilations of cases relating to horses are not presented in such a systematic form as those of human beings.

[22] Hofacker and Ehrhardt took up the figures in the cases numbered 7 to 21 in Notter.

Eighteen months later, the work, newly-printed in Austria, reached Paris. Charles-Chrétien-Henri Marc (1777–1841), one of the editors of the *Annales d'hygiène publique et de médecine légale*, passed it on in the columns of that new periodical, now known as one of the most important early promoters of empirical statistics, despite the obstacles that such research encountered at the Paris Academy of Sciences.[23] This note in French was presented as a translation of the essential points of the German letter, but because in this era there was no clear conception of the cross-tabular structure that governed the headings used, and as a consequence of the complexity and the repetitive nature of the list of 15 cases given by Hofacker, Marc (or his inadequately supervised printer) showed only 14 lines, one of them produced by conflating two originally highlighted by Notter and repeated by Hofacker and Ehrhart: the lines numbered 8 and 9 in the Austrian version were telescoped together. The results were wrong and the categories used to classify parental ages, incoherent.[24]

It will be remembered that *Annales des sciences naturelles* in 1825 and *Annales d'hygiène* in 1829 were both – the first for the physical sciences and the second for medicine – journals newly-established in Paris, signalling a break with the scientific orthodoxy of the Academy of Sciences of that era. Yet, as it passed from one journal to another – from Paris and back again over an interval of four years, via Tübingen and Innsbruck – the structure of the way the two categories of parental ages had been combined became dissipated. It had been very clear to the 1825 editors, and was equally clear two years later in Notter, but it had been transposed into forbidding numbered lists. This "table" was not thought of then as cross-tabulation is now; and although it was constructed on a skilful breakdown of the distribution of both parents' ages around what we nowadays call the modes of their distributions, it was not consolidated in the scientific work then circulating, either as an identified technique or a fortiori as a methodological innovation.

[23] This publication took place later than the reading of Poisson's paper to the Academy of Sciences, but there is nothing to indicate that it arose from a concern to make an immediate reply to that. On the innovative place of the *Annales d'hygiène* in the history of statistics and of social sciences, see Lécuyer (1977a, 1977b).

[24] Thus, Hofacker's Line 8 (1827) corresponded to "young men" and "middle-aged women" – that is, to Notter's Case No 14 (1827, 1828): "fathers aged 24 to 35* and mothers aged 26 to 35*"; his Line 9, to "young men" and "elderly women" – that is, to Notter's Case No 15: "fathers aged 24 to 35* and mothers aged 36 and over". Dr Marc, in his translation, jumped from Line 8 to the text on the following line. Thus he skipped the case of "young men" and "middle-aged women".

Quetelet started from this point when, in 1834, he returned to the question of the proportion of the sexes at birth, faithful to the empiricism of his Paris mentor, Fourier. In both editions of his *Physique sociale [Social physics]* (1835 and 1869), he repeated the unsound numerical results published in the *Annales d'hygiène* in 1829, even leaving in place a wrong pagination (p. 537) that threw his successors and later historians off track. The table he published therefore included not only an error that made it almost incomprehensible, but also a serious flaw: the gradual conjuring away of the empirically-constructed work that had been carried through so attentively from Morel-Vindé (1814) to Notter (1827).

That was not all. Quetelet wanted to make himself heard by scholars all over Europe and, encouraged by Malthus in 1833, wanted to use as an example the innovations in the statistical apparatus of the very young Kingdom of Belgium. In citing Hofacker – not Condorcet, whose published writings he knew perfectly well, nor even any one of Laplace, Fourier or Poisson – he was able to get out of the Paris rut, even perhaps to rid the approach to the issue of its set of religious prejudices. In this regard, the intellectual context in Tübingen and the surrounding area after 1827 was no longer what it had been before 1810. This was because, at the same period – strangely – Hofacker's namesake, Ludwig Hofacker (1798–1828) was preaching a new evangelical radicalism based on the idea that, since Satan was "the supreme *Aufklärer*", the Enlightenment heralded the end of time.[25] Even though they both died in the same year, from the 1830s this anti-rationalist theologian was certainly more celebrated among German-speaking élites than his namesake, the professor of physiology, who was quickly forgotten.

Admittedly, in his *Physique sociale* (1835), Quetelet did mention Poisson's paper, which had appeared in 1830, and Girou de Buzareingues' research. But, in focusing European scholars' attention on conjectures about the influence of parental ages on the probability of a child's sex, he obviously clouded the issue by creating a link between snippets that he attributed to Hofacker and Sadler's conclusions on the English peers. He granted primacy of observation to the former, even though he did not know how it had been established, while he accorded the latter credit for having mobilized figures convincing enough to support his predecessor's initial outline. So, in 1835,

[25] Ludwig Hofacker, *Predigten für alle Sonn- und Festtage*, Stuttgart, Steinkopf Verlag, 1977 (cited here in the 51st edition), Vol. 1, p. 512. See Albert Knapp, *Leben von Ludwig Hofacker*, Heidelberg, Winter, 1872.

this became the accepted record of the genesis of a discovery attributed to "Messrs Hofacker and Sadler" – the kind of speculation about the results of statisticians' work that has been made a hundred times since then to no avail. Quetelet, major predator of figures and reformer of statistics, demonstrated his completely casual attitude to compilation and calculation methods. All this – the way Hofacker's letter was collated in Paris, and its forced pairing with Sadler's figures – allowed the Belgian statistician to circumvent Poisson's verdict. The tale of two parallel discoveries away from the French scene had the effect of encouraging European scholars to reopen the empirical question raised by Condorcet and seemingly encouraged by Fourier. As time passed, this question, reduced to a statistical conjecture born in the England of Anglican reformism and the Germany of neo-Lutheranism and Romanticism, lost any sense of the horizon of progress for a human mind freed from the causes of superstition. Whether as a result of progressive amnesia as the figures made their way around the international circuits, of the enthusiasm of the masterful statistician eager to take a short cut, or even of the positivist hypocrisy that was only to be expected, from then on the question was transformed into a live issue in statistics, appearing to relate purely to the matter of calculation. This evaded three fundamental aspects of the question of regularity in the ratio of the sexes at birth: which philosophy does the possibility of this calculation presuppose – pessimist, optimist, or meliorist? Which intellectual traditions is the calculation based on – theology, calculus of probabilities, or social science? What presuppositions does the calculation conceal, in relation to the animal nature of human beings and to the distinction between domestic animals and animals in general?

Chapter 3

STATISTICAL SOURCES, LAW AND MEDICINE (1846–1876)

1. THE GLOBALIZATION OF FIGURES

The paper published by Poisson in 1830 marks the point from which the history of the calculation of sex ratio at birth and that of the analytical calculus of probabilities, while remaining intimately connected, began to gain a certain degree of autonomy from each other. It is true that, decades later, the mathematician Wilhelm Lexis (1837–1914) and the sociologist Maurice Halbwachs (1877–1945) came to re-examine, each in his own way, the relationship of the concrete regularity of this type of observation to the formal abstraction presupposed by the mathematical analysis of chances (Halbwachs, 1936 [2005]; Lexis, 1876; Stigler, 1986). But, after its rewriting by Poisson, and then for more than a century, the calculus of probabilities was regarded by the most influential French mathematicians as one of the areas of excellence in mathematical analysis and in the theory of functions. Thus, Jules Bienaymé (1796–1878), Joseph Bertrand (1822–1900), Henri Poincaré (1854–1912), Jacques Hadamard (1865–1963), Émile Borel (1871–1956), Henri Lebesgue (1875–1941), Maurice Fréchet (1878–1973) and Paul Lévy (1886–1971) followed a collective trajectory, a kind of "French way" of probabilistic thinking – in terms of intellectual tradition and internal tensions. The history of statistics struggles to reconstruct this tradition because nowadays statistics is dominated by other styles of mathematical thinking, not fashioned long ago in Paris at the *École normale supérieure* or the *École polytechnique* but in the second half of the 19th century and the 20th century in German, British and American universities. Although, with the international circulation of the works of Laplace and of Poisson – and of Carl Friedrich Gauss (1777–1855), their German contemporary and counterpart – the analytical calculus of probabilities began to spread worldwide, it is important to note that Laplace's know-how, his

E. Brian and M. Jaisson, The Descent of Human Sex Ratio at Birth, 59–85.

distinctive contributions, and the very D'Alembertian scepticism that was characteristic of him remained part of an essentially local mathematical culture – whose objects were, it must be said, intensely abstract.

Although this thinking has to be along the lines of a concrete history of abstraction (Perrot, 1992), it is clear that the globalization phenomena in this instance did not operate all of a piece: mathematical concepts have no reason to circulate at the same rate as the figures which carry them along at a given moment in history (Brian, 2001b). In this case, in the wide-ranging debate between Fourier and Poisson, it was the counting processes – not the most abstract mathematical concepts – that saw such a tremendous expansion in the mid-19th century: historians agree in pointing to an explosion of empirical material for calculations (Hacking wrote on this subject in 1990: "an avalanche of numbers"), followed by the ascendancy of new modes of thinking (Porter described it in 1986 as a "rise of statistical thinking").

A mutation like this did not come about by itself – quite the opposite. As a result, on the one hand, of Condorcet's injunctions in the columns of the *Histoire de l'Académie royale des sciences* and in the pages of the *Esquisse d'un tableau historique des progrès de l'esprit humain* (1795) and, on the other hand, of recommendations from Fourier, architect of the *Recherches statistiques sur la ville de Paris...*, the Belgian astronomer Adolphe Quetelet (1796–1874) mobilized scholars and the staff of European statistical offices with a view to standardizing their counting processes. Quetelet took the view that only governments could succeed in gathering the materials needed by the Science that was the object of his ambition (Quetelet, 1853). To this end, between 1853 and 1876, he called about ten sessions of an International Statistical Congress, bringing together the directors of the main specialized statistical offices in Europe and mobilizing about five thousand specialists in all countries, with the sustained twofold objective – administrative and scientific – of improving procedures for recording and investigating the most varied objects. The historical conditions in which this institution consolidated its position demand an analysis that is not part of our current remit (Brian, 1989, 1992, 2002). Among the most obvious factors that encouraged the success of the Congress should be noted the growth in rail links between European capitals, hand in hand with the rapid development of the universal exhibitions movement; the political circumstances in Europe following the 1848 Revolution, which saw alliances forged in many countries between neo-absolutist governments and administrators often inclined towards economic liberalism; the gradual conversion of these governments to tariff-free trade; the processes of political unification in Germany and Italy and the rise of local forms of nationalism – in both cases, factors in the proliferation of new offices of statistics and the reorganization of existing ones.

The literature has been full of historical studies in this area for a good twenty years (Beaud & Prévost, 2000). However, no one has yet produced a rigorous summary with a view to a sociology or an historical epistemology. In the meantime, we can observe – following the statisticians of that era, such as Quetelet himself (1873), Heuschling (1882a, 1882b) or Neumann-Spallart (1886), and the historian Westergaard (1932) – that the Congress changed everything. Its effect was to give administrative counting procedures a minimum of uniformity, without which the tables produced from then on would not have provided so much evidence.

It is also obvious that the second half of the 19th century was a period of great intensity in the circulation of specialized works, and in their translation. This meant that, in Europe, the major statistics textbooks of the late 19th century, on which statisticians were trained from the 1880s to the 1940s – in their own countries and their own languages – were mostly summaries, for local use, of the literature resulting from the Congress (Brian, 1989, 2002).

"I felt then how much it was to be desired that this theory [of probabilities] might be made more elementary, and that it be brought down from the higher reaches of analysis. . .," wrote Quetelet in 1846, in the first lines of one of his works designed to instruct those reading statistics, as well as employees of the administrations that would have to produce them (Quetelet, 1846; see also 1828). In this regard, he was an inventive teacher, continuing the line of Joseph Fourier and Sylvestre Lacroix (1745–1843), himself trained by Condorcet. Quetelet had the ability to translate some of the most tricky elements of the late-18th-century and early-19th-century corpus on probability – as, for example, when he distinguished between and compared the repetition of the same uncertain size measured several times with taking the measurements of several similar things (Armatte, 1995).

We should observe that he reformulated Condorcet's argument on the "probability of future events according to observation of past events" (Condorcet, 1786). But what might have seemed to be, in Condorcet's work, an analytical clarification – the key to which was a mathematical construction of the comparability of successive events – had become, half a century later, an analogical schema that presupposed comparability of events and assimilation between measurement errors and the variability of phenomena (Quetelet, 1846).

For Quetelet, the proliferation of necessarily imprecise measurements for the same known object – he imagined that the same statue was being measured several times – highlighted both the error dispersion and the usefulness of an averaging calculation where the errors were supposed to cancel one other out. Next, when he envisaged using the same process of calculating the average for several similar magnitudes – separate statues would be measured, for example – he stated that one result would always be obtained. In his eyes, this result would be relevant only if the magnitudes measured had derived from the same object, one that was in fact inaccessible – if, for example, the statues had been copied from the same model, now lost. Without this ideal model, the use of the average would be unjustified (Brian, 1991).

In 1846, Quetelet took the view that the edifice of mathematics was already solidly built and that observations, as long as they were properly collected, would lend themselves to these calculations. In 1786, Condorcet had explored mathematical construction and tried to use analytical thinking to describe the repetition of events and the degree of certainty that might be expected from them. By coming down "from the higher reaches of analysis", Quetelet arrived at what we might describe as a positivist paradigm of statistics, where all observations of facts are supposedly established and commensurable. But, in passing, he lost what had given Condorcet's idea its force – the ability to use mathematical construction so that it answered the epistemological question. In Condorcet's work, some passages can be located between his metaphysics of the calculus and the philosophies of Leibniz, Hume and Locke (which he knew) or possibly towards that of Kant (which he did not) (Crampe-Casnabet, 1985); half a century later, in Quetelet's work, the mathematical mechanism was viewed as embedded, and so these routes had disappeared, conjured away by a peculiar syntax that combined error, variability and uncertainty, sealing their relationships and solidifying them. Quetelet's instructions, because they enjoyed tremendous success and became very deeply inscribed in scientific procedures and administrative institutions, caused an epistemological rupture opening what is usually called "positivism" – a rupture that was more decisive than the basic legacy of Auguste Comte's own philosophy.[1]

Quetelet handed down two lasting expressions: "social physics" and "the average man"... Or rather let us be more rigorous and say that, in the social sciences and in statistics, the collective memory of these two notions has focused particularly strongly on the name of this scholar, because the use of the words has made it considerably simpler to transmit the memory of the enormous, intense collective activity framed by international meetings of statisticians in the mid-19th century. To speak of "social physics" would be to maintain the myth that, in the first place, the mathematical and physical sciences had developed up to the early decades of that century and then, once they were established, a physics of social phenomena had followed from them. If we think of the research of scholars, nothing could be further from the truth, as we have already noted above in relation to the calculation of the proportion of the sexes at birth. The 18th century was a period of very serious enquiry into social, political and economic phenomena, admittedly by routes that could be described as "pre-disciplinary" (Heilbron, 1990; Heilbron, Magnusson, & Wittrock, 1998). The reform of mathematical analysis and the formulation of the analytical calculus of probabilities in the late 18th century embraced issues of physics just as much as those of morality or of politics (Brian, 1994a). And Quetelet's slogan expressed one particular option in a panoply of new concepts in the moral and political sciences, an option characterized in this case by a form of division of administrative and scientific labour, in which observation was a matter for government offices and mathematical calculations, for learned societies (Brian, 2001a, pp. ix–xv and 2002).

The apparent simplicity of the notion of "the average man" has had the retrospective effect of blurring the principle. The average man as an ideal model provided the foundation for use of the technique of arithmetic means: each human being was seen as something approaching a copy of the average, and the regularity of distribution of variations in height and other measures obtained by recording statistics most often resembled the regularity of the error distribution (Armatte, 1995; Stigler, 1986). For Quetelet, the construction and harmonization of administrative statistics, discussed throughout the sessions of the International Congress,

[1] This rupture throws light on the profound lack of understanding that has remained between the legacy of Quetelet's positivism and the mathematical tradition from Poisson to Fréchet and Lévy, which we have already outlined.

were directed towards highlighting this ideal. They opened the way towards building a comparative international statistics (Heuschling, 1882a; Quetelet, 1873; Quetelet & Heuschling, 1865). Large-scale reform of offices of statistics aimed to establish empirical observation on the basis of the criteria of a now-simplified calculus of probabilities:

> "The calculus of probabilities is merely the instrument that must serve to regularize operations; but it is becoming indispensable in the researches to which we must devote ourselves. It should help us to distribute the series of our observations to advantage, to estimate the value of the documents we shall be using, to distinguish those that exert more influence, then to combine them in such a way that they deviate as little as possible from the truth, and eventually to calculate the degree of confidence that can be attached to the results obtained. The theory of probabilities basically teaches us only to do with more regularity and precision what even the wisest minds have done somewhat vaguely up to now. Above all, in the phenomena that we shall have to deal with, it aims to substitute science for what has been called practice or experience, and which, most of the time, is merely blind routine." (Quetelet, 1846, p. 7).

Laplace concluded his works with a famous Cartesian recollection: "the theory of probabilities is at bottom only common sense reduced to calculus" (1812, p. CLIII of 1886 edition; 1814, p. 105 of Vol. 2 of 1921 edition; translation by Truscott and Emory from Laplace, 1951, p. 196). As for Quetelet, he privileged one aspect of this calculus, abandoning both his theory and common sense on the heights of analysis, in order to promote a bureaucratic revolution that consisted of giving the simplified calculus primary importance in the administrative setting. Admittedly, in the same works, Laplace had also pleaded that it was "very important to keep in each branch of the public administration an exact register of the results which the various means used have produced, and which are so many experiences made on a large scale by governments." (1812, p. LXXVIII of the 1886 edition; 1814, pp. 1–2 of Vol. 2 of the 1921 edition; translation by Truscott and Emory from Laplace, 1951, p. 107). But this mark of Enlightenment idealism, which by thirty years later had become a pragmatic international objective, was open to a multitude of real constraints. Thus, in the second half of the 19th century, offices of statistics were co-ordinated into a vast international laboratory, deliberately designed – from the level of bureaucratic organization to the training of clerks in the required know-how – to serve a particular scientific vision: a vision simplified by Quetelet himself and exploited by the directors of these offices, who were very active in the Congress (Brian, 2002).

In fact, neither his colleagues nor his successors invariably subscribed to the founder's concept, although they served it very effectively.[2] Starting

[2] Quetelet was a victim of his own success even within the Congress, as this exchange, on Wednesday 8 September 1869 in The Hague (CIS, 1869, Vol. II, pp. 69–70) indicates: two pillars of many sessions of the Congress, the Prussian Ernst Engel (1821–1896) and the Englishman William Farr (1807–1883), drove him into a corner – the former out of clumsiness, the latter not without humour:

Quetelet. – [...] Although it is thought that a man is born, grows up and develops according to chance, there is actually the most regular law presiding over his growth. [...] I repeat, the law of his growth is extremely regular; but, to discover this law, a large number of observations must be assumed. [...] Take the moral qualities of humankind, and you again encounter this same law. [...] This, I repeat, applies not only to a man's physical development, to height, weight, strength; but even to the crimes that are committed. I have demonstrated this in my books[*1] [...].
Engel. – Monsieur Quetelet has just said that the discovery of the law of the development of human size is something new. I must remark that Polyclitus[*2] discovered this law over two thousand years ago. (Hilarity)[*3].
Quetelet. – I am not claiming to have discovered it. *Non equidem invideo, miror magis...*[*4].
Engel. – It was Monsieur Chateau, who, in 1811[*5], rediscovered Polyclitus' work on man; this work indicates all the dimensions of human size from youth to old age. He measured these dimensions on a large number of individuals, and established the average [...].
Quetelet. – Before publishing my first book on the size of man and his development, I made a study of all that had been written about man[*6]. First of all, I must point out to you that the theory of averages was not known to the Greeks.
Engel. – Polyclitus did not make up an algebraic formula; he established his dimensions by experimenting.
Quetelet. – The work you are talking about may be a curiosity, but there are a lot of copyists' errors in it. Moreover, the law we know nowadays, which is based on the theory of averages, is not to be found there.
Engel. – I am not saying that the distinction of having made the discovery of this law is due to Polyclitus; but I am saying that people were looking at the sizes of man over two thousand years ago, and that the most famous artists looked at the matter, as the immortal sculptures of Greece prove[*7].
Quetelet. – They did not know of any law.
Engel. – No, but they arrived at perfection by instinct.
Farr. – Has Monsieur Quetelet measured the principal statues of Antiquity?
Quetelet. – Yes. And I must tell you what happened with the statues [...][*8].

[*1] Quetelet is here putting forward the culmination of all his works, which had just appeared – the revised second edition of his *Physique sociale* (1869).
[*2] Polyclitus: Athenian sculptor, 5th century BCE.
[*3] The minutes also record reactions in the hall.
[*4] Virgile, Bucolics I, 11: "I am not jealous, but rather, amazed...".
[*5] The stenographer and the editor of the minutes were unable to reproduce the name of Gottfried Schadow (1764–1850), sculptor of the quadriga on the Brandenburg Gate and author of *Polyclète*

soon after his death, they made a rapid re-examination of the profusion of numerical materials, approaching it by other routes. For example, Louis-Adolphe Bertillon (1821–1883), director of the Office of Statistics for the City of Paris, went back to the same thinking on averages; however, he no longer started from an average man but from several types – all also ideals, but distinct ones. This became one of the most persistent empirical arguments of racial anthropometry (Brian, 1991).

2. THE LEGEND OF THE "HOFACKER-SADLER LAW"

In our eyes, the balance between an averaging process and the concept of the average man is a *petitio principii*, in the same way as the balance between anthropometry and the race hypothesis: it is important not to merge the calculation technique with its favoured object. In effect, Quetelet's body of scientific work, his publications and his activism in international meetings brought the two together. As a young man, he was not so dogmatic. Closer in this regard to Joseph Fourier (1821/1829) and to doctors working in public health, like Louis-René Villermé (1782–1863), but also to Denis Poisson (1830) or even the young Laplace (1785/1786), he asserted that the calculation of averages and of differences from averages enabled analysis of causes (Quetelet, 1832). This way was a quasi-experimental approach, where the average was a trial rather than a norm. A major part of his research focused on the study of periodic variations in human phenomena and, in particular, of their analogies with observable meteorological fluctuations.[3] Thus the average man was only one of his results – though probably the most glorious.

ou théorie des mesures de l'homme selon le sexe et l'âge, Berlin, [1834] (new edition, 2 vols. bilingual, Berlin, Amsler and Ruthardt, 1866–1867). This neoclassical artist wrote his book in the context of Romanticism, drawing on early 19th-century physiognomy. It includes an essay on the diversity of human heads. Its reissue came in different circumstances, at a time of the rapid development of anthropometry.

*6 Here he touches on the first edition of *Physique sociale* (1835). No edition of *Physique sociale* mentions either Schadow or Polyclitus. It must be freely admitted that Quetelet was right – the Prussian sculptor was not immune to anachronisms.

*7 It will be remembered that Quetelet's theory of errors rests on an apologue in which statues occupy an important place. Ernst Engel's tactless flattery in fact touches on Quetelet's Platonism.

*8 He then goes on to state that the statues of Antiquity came from an assemblage of elements taken separately from different models.

3 This was without doubt a result of his reading of the "important remark" that concluded Laplace's work (see Note 3, p. 32). Thus, when viewed over the long term, empirical meteorology and the mathematical theory of fluids have profound connections with the history of statistics and the history of the calculus of probabilities.

Quetelet tackled the issue of the proportion of the sexes at birth with a fairly open mind, taking up Laplace's conclusions on the excess of male births (Quetelet, 1846) but mapping a route towards deeper empirical research into the causes (Quetelet, 1835, 1869). This route accorded with Condorcet's recommendations in the passages of the *Esquisse* (1795) on studying the transmission of physiological and moral qualities from one generation to the next. Compiling earlier research for the first edition of *Sur l'Homme et le développement de ses facultés, ou essai de Physique sociale [On Man and the development of his faculties, or an essay in Social physics]* (1835) gave Quetelet the opportunity to put forward a conjecture about sex ratio at birth: that this proportion might depend on parental age gap at the time of conception. As already mentioned, in doing this, he combined two names in one expression for the first time – "Hofacker and Sadler" (see above, p. 57). The fact is that, for a century after him, anyone who wanted to tackle the issue of sex ratio had to surmount what was frequently referred to as the "Hofacker-Sadler law" (Gini, 1908; Halbwachs, 1933, 1936; Legoyt, 1864–1870; Stieda, 1875). This conjecture created a third legend from a mixture of prejudices and partial results. Two parallel empirical discoveries around 1829–1830 – one in Germany, the other in England, combined by a young Belgian scholar in 1835 – may be said to have dragged the examination of causes of enigmatic regularity away from the mathematical debate being conducted in France in that era at the "higher reaches of analysis". Contemporary specialists could have known that, in England, Sadler was an anti-liberal parliamentarian revisiting – along the same lines as Pastor Johann Peter Süssmilch – the laws of population proclaimed by the Reverend Thomas Robert Malthus (1766–1834) in his critique of Condorcet's *Esquisse* (Malthus, 1798). Those with a close interest in issues of religion in Germany could easily have confused the name of the enlightened doctor and Francophile Johann Daniel Hofacker with that of Pastor Ludwig Hofacker (1798–1828), a mystic then famed for his neo-Lutheran criticism of the Enlightenment. Whatever the case, the "Hofacker-Sadler" myth had the effect of uprooting investigation of the sex ratio from its context – Condorcet's anti-superstitious thinking set against the background of the French Revolution.

This movement blurred the philosophical and metaphysical markers of the end of the previous century; it did so very particularly at the end of a volume of minutes of the first session of the Statistical Congress, held in Brussels in 1853, in a "Paper on the philosophy of statistics", written by Joannès-Erhard Valentin-Smith (1796–1891), statistician and senior magistrate in Lyons (CIS, 1853, pp. 239–254).

"When man knew his future destiny, that day would bring turmoil to all laws, divine and human. From that day on, man would no longer have any willpower or free will; he would no longer be anything more than a sort of thinking automaton, sliding about, lost in the machinery of a life entirely functional and mechanical. Through the great goodness of Providence, what we are fortunate enough to be able to anticipate in the general sphere of the masses, we escape in the limited sphere of individuals; what the science of statistics indicates is a law certain in the land, or even in the family of the community, cannot be discerned in the domestic setting. What a wonderful effect of divine wisdom, which alongside certainty has placed uncertainty, alongside light, darkness, thereby teaching that, for the very happiness of man, everything must be at the same time revelation and mystery in the forward march of humanity." (CIS, 1853, p. 242).

Here, the aspects of the scientific intentions of Süssmilch and of Laplace that were diametrically opposed are deliberately merged.[4] Laplace's general cause – the "facility of births of boys", which in his eyes was the equivalent of D'Alembert's general cause of the winds – was transformed into an intention from on high. Did Quetelet himself share these views? His speech

[4] Let us summon Laplace: "Events which depend on chance offer as a whole a regularity which appears to hold a design, but which is, at bottom, *only the development of their respective possibilities*. The ratio of births of boys to that of girls, in the great cities such as Paris and London, is an example of this; this ratio varies very little: some scholars think they have seen in this constancy a proof of the Providence that governs the world; but the analysis of chances shows us that this ratio must always nearly coincide with that of the *facilities of birth* of the two sexes." (1795, p. 131 of the 1992 edition). The terms that we have put into italics express mathematical concepts present in Laplace's early research: the "development of their respective possibilities" is the integral of the equation of the phenomenon, taken for all possible values of the variable that defines the possibilities; the "facility of births" is the *a priori* probability of a type of birth under consideration. This passage was to be slightly altered by the author and used again in *Théorie analytique* and in *Essai philosophique*, and this time it was he who added the emphasis to the word *chance*: "Amid the variable and unknown causes which we comprehend under the name of *chance*, and which render uncertain and irregular the march of events, we see appearing, in the measure that they multiply, a striking regularity which seems to hold a design and which has been considered as proof of Providence. But in reflecting upon this we soon recognize that this regularity is only the development of the respective possibilities of simple events which ought to present themselves more often when they are more probable." (1812, p. XLVII of the 1886 edition; 1814, pp. 55–56 of Vol. 1 of the 1921 edition; translation from Truscott and Emory, 1951, p. 60). This time he expresses himself even more rigorously. Nothing here lends itself to Valentin-Smith's vision.

to the seventh session of the Congress in The Hague, in defence of *Physique sociale* at the time of publication of its second edition, raises doubts.

> "It seems to be believed that man was flung into the world without any law: it has been said that he obeys his own free will; and that's all there is to it. This a profound error. If you take a large number of individuals, here again you encounter the most regular of laws." (CIS, 1869, Vol. II, p. 69).

But here again, in comparing Laplace and Quetelet, there is another noticeable reversal. Although the former challenged physico-theology, the latter seems to have been looking at a simplistic notion of free will, where humankind behaves erratically. Of prime importance to Quetelet as an astronomer was a concept of regularity where error, fluctuation and uncertainty were alike, and should be reduced in order to reveal a primary regularity. At this point, it mattered little whether this regularity was physical or theological... This simplification, moreover, completely suited the architects of the internationalization of statistics, whose first concern was not philosophical clarification. In order to reconcile conservative élites who set store by religion and other, often more progressive, liberals, it was better to give priority to the issues involved in organizing offices and circulating results. This led to the creation of a semi-academic vulgate. No longer leaving room for either the skilful scepticism of a D'Alembert or the radical conjectures of a Condorcet, it mixed a semblance of Laplacian philosophy with a hotchpotch of theology – vestiges of Süssmilch's old Leibnizianism with signs of the recent revivals of forms of mysticism that had taken place in all Christian persuasions. Thus, statistical positivism adjusted the philosophical and political legacy of the Enlightenment to the mysteries of faith.

3. THE MAKING OF BIRTH RECORDS

But what of the matter of counting the sexes at birth? In the eyes of Congress participants, this meant improving birth records. By then, two different situations pertained in Europe. Either these lists were maintained by the religious authorities, as in Orthodox Russia, Catholic Austria or Lutheran Sweden. Or else they derived from civil status in its true sense – that is, registration procedures were regulated by the local administrative authorities on the basis of the Napoleonic Civil Code; this, first drawn up in March 1804 towards the end of the Consulate and then revised in 1806 under the Empire, spread farther afield as the century went by (see Box on p. 70).

Civil Code of Year XII (1804)

Of Acts of Birth. Art. 55. Declarations of birth shall be made, within three days after delivery, to the civil officer of the place: the child shall be shown to him.

Art. 56. The birth of the child shall be declared by the father, or, in his default, by the doctors in physic or surgery, the midwives, the officers of health, or other persons who shall have attended at the birth; and where the mother shall have been delivered out of her own house, by the party at whose house such delivery took place. The act of birth shall be immediately reduced to writing, in the presence of two witnesses.

Art. 57. The act of birth shall set forth the day, the hour, and the place of birth, the sex of the infant, and the *forenames* which shall be given it, the *forenames* and surnames, profession, and domicile of the parents, and those of the witnesses. [...]

Of Successions. Art. 718. Successions are opened by natural death and by civil death. [...]

Art. 720. If several persons, respectively called to the succession of each other, perish by one and the same accident, so that it is not possible to ascertain which of them died first, the presumption of survivorship is determined by the circumstances of the event, and in defect of such, by force of age and sex.

Art. 721. If those who perished together were under fifteen years, the eldest shall be presumed to have survived. If they were all above sixty, the youngest shall be presumed to have survived. If some were under fifteen years, and others more than sixty, the former shall be presumed to have survived. [...]

Art. 723. The law regulates the order of succeeding between legitimate heirs: in defect of such, the property passes to natural children, afterwards to the father or mother surviving; and if there be neither of those, to the *Republic**. [...]

Art. 725. In order to succeed, the party must of necessity be in existence at the moment at which the succession is opened. Those incapable of succeeding are 1st. He who is not yet conceived; 2d. The child who is not born likely to live; 3d. He who is civilly dead. [...]

Art. 731. Successions are decreed to the children and descendants of the deceased, to his ancestors and collateral relations, in the order and according to the rules hereafter determined. [...]

Source: Code civil des Français. Edition originale et seule officielle [French Civil Code. Original sole edition], Paris, Imprimerie de la République, Year XII (1804); based on 1827 translation, attributed to George Spence: the italicized words indicate our revisions to this historic translation.

* Here the 1806 Code refers to "the state" and not "the Republic".

Twelve years after the first meeting of the Congress, Quetelet and one of his Belgian colleagues published, in Brussels, a *Statistique internationale (population) [International Statistics of Population]* (1865), designed to serve as a prototype for the comparative publications encouraged in the course of the sessions, for which directors of statistical offices in the main countries had provided previously unpublished materials. In this work, Quetelet deplored the lack of uniformity of registration documents, and implied that he viewed the degree to which the Napoleonic Code was being applied as the measure of the progress of Science:

> "Countries may be found, for example, even among the most enlightened ones, where the recording of births remains alien to the civil authorities, and is done by ministers of religion: the result of this is that the number of children recorded could, owing to omissions, be lower than in reality." (p. XXXII).

A textbook case is provided by the comparison of registrations of stillborn infants, which was then particularly difficult: any count of stillbirths was open to doubt, and it was unknown whether they appeared in the different countries in the same way, or in fact were shown in one place as births, in another as deaths, or elsewhere as both – even if they showed an excess of boys (p. XXXII).

The study was therefore condemned to be limited solely to children born alive. This was the opportunity for Quetelet to revise Poisson's assertion of 1830: with the proportion of the sex ratio at birth, we are entering the realm of a *social* phenomenon. However, he abandoned his predecessor's mathematical rigorism.

In order to avoid anachronism, we should clarify that the legitimacy of a birth (although it is a "social fact" in the later meaning from Durkheimian sociology) is here not *social* in the sociological sense of the word. Nor is it *social* in Condorcet's sense of a *social art*, which would be in the nature of a reasoned use of the moral and political sciences in matters of public concern – relating even to the use of the sciences in general. For Poisson in 1830 and for Quetelet in 1865, the description *social* pertained to the institution of the law as much as to moral judgement: *illegitimacy* is an issue of immorality, of the clandestine and of a relative lack of forethought – all distinctive causes that are disturbing in the eyes of these authors.

Using recent methods that accord with those of 1830 (see Table 2, p. 72), we can confirm that, according to figures published by Quetelet

Table 2: Quetelet's figures (1865)[5]

Country	Years	Total	Legitimate live births	
			Sex ratio	CI95%**
Greece	1861	32 174	51.77%	0.56%
Spain	1858–1861	2 160 076	51.68%	0.07%
Hanover	1854–1858	256 957	51.56%	0.20%
Kingdom of Saxony	1859–1861	232 678	51.50%	0.21%
France	1851–1860	8 830 186	51.32%	0.03%
Prussia*	1859–1861	2 016 240	51.42%	0.07%
Bavaria	1851–1860	1 191 290	51.39%	0.09%
Netherlands	1850–1859	1 030 517	51.34%	0.10%
Belgium	1851–1860	1 262 743	51.32%	0.09%
Austria	1854–1857	5 048 190	51.30%	0.04%
Norway	1851–1860	449 241	51.27%	0.15%
Sweden	1856–1860	573 004	51.21%	0.13%
Russia	1858	2 776 381	51.19%	0.06%
England*	1860	640 355	51.17%	0.12%
All	about 1855	26 500 032	51.34%	0.02%
Country	**Years**	**Total**	**Illegitimate live births**	
			Sex ratio	CI95%**
Greece	1861	231	51.52%	6.58%
Spain	1858–1861	127 467	51.01%	0.28%
Hanover	1854–1858	29 267	50.89%	0.58%
Kingdom of Saxony	1859–1861	42 355	50.94%	0.49%

[5] In this table "England*" stands for "England and Wales". For Prussia*, stillbirths are included in these birth figures. Finally the column "CI95%**" shows the half amplitude of a 95% confidence interval according to a binomial test whose principle is shown below. For instance, it should be read as: the probability resulting from observations that the *a priori* probability (in Laplace's terms, the facility) among legitimate births of the birth of a boy will be less than 51.32% (*i.e.*, 51.34% − 0.02%) or greater than 51.36% (*i.e.*, 51.34% + 0.02%) is under 5%; and that of its equivalent for illegitimate births being less than 50.03% (*i.e.*, 51.00% − 0.07%) or greater than 51.07% (*i.e.*, 51.00% + 0.07%) is under 5%. This means that the frequency measured in each of the two cases also has slim chances of being the value of the *a priori* probability of the other case. See Appendix D.

France	1851–1860	705 747	50.82%	0.12%
Prussia*	1859–1861	184 053	51.26%	0.23%
Bavaria	1851–1860	332 250	50.95%	0.17%
Netherlands	1850–1859	45 462	50.82%	0.47%
Belgium	1851–1860	108 454	50.62%	0.30%
Austria	1854–1857	490 934	51.28%	0.14%
Norway	1851–1860	43 060	51.21%	0.48%
Sweden	1856–1860	55 233	50.52%	0.43%
Russia	1858	120 591	51.34%	0.29%
England*	1860	43 693	50.72%	0.48%
All	about 1855	2 328 797	51.00%	0.07%

Source: Quetelet and Heuschling, 1865, pp. xxxvii and xxxix.

in 1865, it is highly unlikely that legitimate and illegitimate live births in mid-19th-century Europe included the same proportions of the sexes. Quetelet – repeating the assertions of Laplace and of Poisson in a passage that he used again word for word four years later in the second edition of *Physique sociale* – conjectured strongly on a physiological cause. He wanted to explain the excesses of boys by following in the footsteps of a Berlin public health specialist who had done work on premature death, Johann Ludwig Casper (1796–1864), and in the footsteps of the theory of generation emanating from none other than Girou de Buzareingues, the Avreyon agronomist whose theory Poisson had dismissed forty years earlier.

"There is a curious thing about this table: the excess of male births over female births can be observed in all the countries subjected to our calculations; but this excess is more marked for legitimate births than for illegitimate births: in general, there is no country that is an exception on this point. It is natural to seek the explanation for this difference: since the legitimacy of a birth is the result of a social act, the difference noted must be attributed to the care lavished on the child and on the mother during delivery. But, all things being equal, if more care is given to legitimate children, fewer children will be lost at the moment of birth, and especially fewer boys, whose arrival is surrounded by more dangers than that of girls. [...]. In general, it can be said that the less care is provided at deliveries, the more male children must be lost proportionally." (Quetelet & Heuschling, 1865, p. XXXVII; Quetelet, 1869, pp. 241–242).

"If we were desirous of guessing at this point, we might say, with those who suppose that a male conception requires a certain excess of energy in the woman, that this excess of energy was absent or wanting during the growth of the foetus, and that energy failing, the child would suffer more from it, if a boy, than a girl. Hence the disproportion of dead births between the two sexes; hence, also, the greater mortality of boys immediately after birth, and during the period of suckling, at which time they are still in some measure connected with the mother." (Quetelet, 1869, pp. 223–224; 1842 translation of 1835).

In scrutinizing still-births and pleading the cause for more precise registration, the founder of the Congress reached the limits of his work and of using the whole of humanity as a laboratory. As we have seen, his 1865 criticism of weaknesses in the collection of birth data implied that registrations arising from the French Civil Code offered the best model, and he aimed to extend its provisions to other countries, just as, in other sessions, the Congress promoted the metric system. Things had already been moving in that direction: at the first two sessions in Brussels (1853) and Paris (1855), the recommendations agreed by directors of European offices of statistics enjoined them to bring international pressure to bear on their respective governments.

But after almost fifteen years, with preparations for the seventh session in The Hague in 1869, the impotence of such wishes had to be acknowledged, and there had to be a change of strategy. The Congress had come up against the stumbling-block of the legal and practical reality of registration, which had never been an *ab initio* measure designed for sophisticated use in the sciences, as Condorcet had hoped for, but always a product of a combination of rules laid down over a long period (whether arising from the law or not) with actual concrete practices that were adjusted at the convenience of registration officials and according to the necessities of preserving written records. Making entries in a register – given that here registration is an act taking place in civil law – transmutes this organized jumble of rules, traces and dust into legitimate marks which specialized officials select and then translate in support of certificates attesting the existence of living or dead people, documents drawn up in order to help to resolve disputes and conflicts between the living (Cerutti & Pomata, 2001; Vismann, 2001).

A report by two Dutch public health statisticians on this topic formed part of the programme; they declared not only that registrations of births were heterogeneous within the same country, whether governed by the Code or not, but also dismissed unfavourably both ecclesiastical procedures, which were not very reliable, and the Napoleonic legislation, which was unsuited to the purpose. It was indeed true that the application of the Code and of rules derived from it blurred the distinction between stillborn children, those who died during delivery, and even live births followed by death before the end of the legal declaration period of a few days. This opened a Pandora's box (see Box entitled "Still-born children registration under scrutiny" on pp. 76–77).[6]

The head of the General Office of Statistics for France, Alfred Legoyt (1812–1885), reacted by asserting both the validity of the criticisms and the impossibility of going against the law. He even outlined an analysis of the attachment felt by the civil authorities and by families to the Civil Code: it resulted from the fact that, as a consequence of other articles of the same Code, official declarations of births fuelled inheritance disputes (see Box entitled "Civil Code of Year XII", on p. 70).

[6] The editors of the report presented in the Congress programme were Johannes Adrianus Boogaard (1823–1877), Professor of Medicine at the University of Leiden, and Lucas Jacob Egeling (1824–1892), Inspector with The Hague Health Service and a specialist on the issue of prostitution, who were both delegates of the Dutch Society for the Progress of Medical Sciences. Several people who regularly attended the Congress spoke during the discussion, almost all with an official mandate from their governments: Giovanni Anziani, section head at the Italian Directorate of Statistics in Florence; Matthaeus von Baumhauer, Director of the General Statistical Division of the Netherlands and President of the Congress in The Hague; Fredrik Theodor Berg, Chief Medical Officer of the Swedish Central Office of Statistics; Pavel Blaramberg, a young statistician from the St Petersburg Office (who was also a composer); Christian Nathan David, Director of the Danish Office of Statistics and a former government minister; William Donnelly, Registrar-General for Ireland; Ernst Engel, Director of the Prussian Office of Statistics; William Farr, Superintendent of Statistics in the Office of the Registrar-General for England and Wales; Peter F. Rudolf Faull, Director of the Office of Statistics for Mecklenburg-Schwerin; Friedrich Hardeck, Director of the Office of Statistics for Karlsruhe; Xavier Heuschling, Secretary of the Belgian Central Commission for Statistics; János Hunfalvy, a physician and Professor at the Polytechnic Institute in Pest, Hungary; Vladimir Jakchitch, Director of Official Statistics for Serbia; Anders Nicolai Kiaer, Director of the Norwegian Office of Statistics; Alfred Legoyt, head of the General Office of Statistics for France; Alexander Mansolas, Director of the Greek Office of Statistics, Gustav von Mayr, Director of the Office of Statistics for the Kingdom of Bavaria; Alexandru Pencovici, Director of the Statistical Division of the Romanian Ministry of the Interior; Louis Theodore Petermann, Director of the Royal Statistical Office of Saxony; Pyotr Semenov, Director of the Russian Central Committee for Statistics; and of course Adolphe Quetelet, President of the Belgian Central Commission for Statistics and founder of the Congress. Others were official representatives of more humble institutions, such as Claude-Étienne Bourdin, a senior French doctor and delegate of the Paris Statistical Society, or Eugène Janssens, a doctor who represented the City of Brussels. The only person to speak without a particular mandate was the Estonian statistician, Ernst Kluge (CIS, 1869, Vol. I, pp. 32–34 and Vol. II, pp. 71–92).

Still-born children registration under scrutiny (1869)

Boogaard and Egeling's assessment (Netherlands). It cannot be denied that in the statistical data on population, notably in those on births and deaths, we encounter a disturbing element that, far from being able to be disregarded, seems only to contribute to doubts about the possibility of ever obtaining a perfect and accurate knowledge of ratios of births and deaths. This element is referred to as *stillbirth.* Great diversity in the registration of stillbirths in different countries and frequent lack of uniformity in registration within the same country distort the relationships between births, stillbirths and deaths. While this confusion is not removed, we shall never achieve satisfactory results.

In [the] registers [of countries governed by the Napoleonic Civil Code] it is not *permitted* to make any distinction between *true stillbirths*, that is to say infants who have died before or during birth, and infants who, *born alive*, have died before the birth is declared to the registration official; a declaration that must take place in the three days after the birth.

In [other countries] like Austria, civil law takes the view that any child whose death is not proved at the time of birth is *born alive.* The proportionate number of stillbirths in Austria is so minimal that Professor Wappäus, in his book on the statistics of the population, has already expressed doubts about the accuracy of the official data.

The legal provisions, which vary in different countries, must be viewed as the greatest obstacle. The law, whether in the social interest or in order to avoid any dispute on the question of whether the infant was living or not, looks at the definite date presented by the entry in the register or by the birth certificate; the doctor carrying out the post-mortem examination, in the interest of science, looks for signs of life after the child comes out of the mother's womb. How are the social interest and the scientific interest to be reconciled?

Legoyt (France). I have personally tried to get registration officials to note whether or not the child was born alive. Some mayors have acted on the invitation from the Minister to make this distinction; but others have protested vigorously and maintained that what they were being asked to do was illegal. These same mayors appealed to the Minister of Justice and complained that his colleague, the Minister of Agriculture and Trade [then responsible for the Office of Statistics], was ordering them to make a special entry in the record that was not only not within the law, but was against the law. The 1806 Decree was inspired by considerations of civil law. The desire was to anticipate the very serious fact that issues of survival, which would include inheritance issues, were decided on the basis of declarations by a doctor or a midwife that were sometimes inaccurate and sometimes self-interested. [...] I recognize that, from the point of view of statistics, there is a very great awkwardness there, because we are not able to recognize which are the children who were living and thus should appear first in the birth figures and subsequently in the death figures. I recognize that, from this point of view, international comparison is completely impossible.

Bourdin (France). There are examples of doctors who have been called to gatherings of men and women who have the air of forming a family in the civil sense, where in fact no marriage has taken place. These doctors did not make any declaration and were condemned for it. I know a doctor who was prosecuted for something like this; he was convicted, and very unjustly convicted; for he was a very worthy man who had been misled.

Hunfalvy (Hungary). It is not possible to state exactly the number of stillbirths in Austria and in Hungary. Registers of births, deaths and marriages are in the hands of the clergy, who make precise entries in the registers only concerning infants who have been baptized. The baptism ordinarily takes place four or five days, sometimes even a fortnight, after the birth, and infants who die during this interval are often not registered.

Kluge (Estonia). I would like to direct your attention to another question, that of infants found dead and their registration. Almost all countries have a large number of infanticides. The dead bodies of these infants are thrown away or hidden in places from which they are generally not recovered until several weeks have passed, or even, especially in the countries of Northern and Eastern Europe, several months, when the thaws of early spring arrive. – Should these infants be recorded as stillborn or as live births? In several countries, they represent a very large addition to the figures for stillbirths. They form up to 20%, and even, during the exceptional war years of 1854–1856, up to 50% of stillbirths in Reval, the governmental capital of Estonia.

Jakchitch (Serbia). Our registers of births, deaths and marriages are very well maintained. However, there is no heading for stillbirths to be found there, [for] the good reason that there are none in our country (*interruption and exclamations*).

Donnelly (Ireland). In Ireland, as in England, stillbirths are not registered.

Kluge (Estonia). I shall take [...] the liberty of adding that stillbirths are not by any means registered everywhere with the same accuracy. Entries in the registers in towns are much fuller than in the country.

Source: Minutes of the Congress (CIS, 1869, Vol. I, pp. 32–34 and Vol. II, pp. 71–92).

It is true that the anonymous author of the *Mémoire historique et instructif sur l'Hospice de la maternité [A historic and instructive paper on the Maternity Hospital]* (1808, p. 25), already cited, deplored the fact that such reasons motivated families in Parisian polite society to lay siege to the Foundlings Hospital with a view to discreetly adopting an abandoned newborn baby. Similarly, confirmation of what Legoyt said can be found in the comments of the jurist Georges-Antoine Chabot (1758–1819) on the right of succession (1818). When, for example, both mother and child die during the course of a delivery, the order of the deaths is important. If the mother dies first, the child inherits, even if it is just for a few moments; then the father and any possible siblings inherit from the stillborn child. If the child dies first, no question of succession arises. Then when the mother dies, her fortune and her dowry go to the mother's parents and not to her husband. The declaration before the civil registration official, which fell to the father and was confirmed by witnesses whom he had called (specifically, almost always other men, as identified in the registration documents themselves), had the effect of throwing a highly official veil over all that had happened between the beginning of the delivery and the time any approach was made to the administrative authorities – frequently, several days. Thus two sections of the Civil Code combined to create a favourably grey area in the regulation of family interests.

As we can see, issues of inheritance in families – most especially in those endowed with commercial or industrial goods or property, both in the city and the country – governed the implementation of the Code's articles on births. The Code reserved to fathers of families and to elected municipal representatives – that is, those with a primary interest – the right to register newborn infants as living or dead and to control this right; so it should come as no surprise that these registers were not able to provide scientifically rigorous documentation. Legoyt had no difficulty recognizing that he could do nothing about this.

Having come up against this limitation, discussion in The Hague then turned to a review of the procedures applied in different countries. This was the opportunity for a great spate of revelations, as participants, spurred on by national vanities, tried to outdo one another; these revelations confirmed in passing the significant extent of infanticide. Moreover, some statisticians showed themselves to be especially dazzled by their instrument. One director of an office of statistics asserted, risking protests from his counterparts, that "there are no stillbirths in our country". Another, more cautious, conceded that his country ignored them. A doctor attested that one of his colleagues had been prosecuted for having refused to record a birth that was evidently taking place outside marriage.[7] This speaker appeared to hold particularly rigid views, and other statisticians of a more empirical bent were reserved in their reception of his declaration that:

> "I am saying that the products of conception are considered to be a child when you are able to distinguish between the sexes. There you have a definite starting-point. When there is no positive distinction between the sexes, there will always be products of conception, but products that should not figure in the register" (CIS, 1869, Vol. II, p. 76).

Sustained by an intellectual tradition going back a century, those promoting general statistics viewed ecclesiastical registrations as too opaque; however, the members of the 1869 Congress in The Hague were forced to record that registration governed by the Civil Code was no more satisfactory when it came to the true registration of biological births. The Code did not provide the support of a rigorous definition of the physics of the

[7] Thus, we can counter an analysis that accords a mysterious power to the text of the Civil Code, claiming to make the era of the Code incommensurate with the previous one, and at the same time to give an advantage to infanticides; instead we can assert the relevance of investigating changes in the social division of labour for producing symbolic instruments, two typical instances of which are here differentiated by the application and the non-application of the Civil Code. Further confirmation of this can be found in French 18th-century works that show traces of the same phenomena, but in a different configuration: this time the specialists concerned in the conflict are priests, midwives, surgeons and physicians. Marguerite du Tertre de La Marche (1638–1706), chief midwife at the Paris Hospital, devoted several pages of her *Instruction* (1677) to the fact that her pupils could find themselves having to baptize a child "when [they] doubt that they will be able to carry the infant to church because of its weakness" (pp. 101–105, 1710 edition). The same determination to administer this sacrament quickly in case of difficulties – and therefore to organize the setting for delivery accordingly – was expressed by master surgeon Pierre Dionis (1643–1718) in his *Traité des accouchements* (1718, p. 312). The full significance of his skills was discussed even further: "It is for the surgeon to neglect nothing to discover whether the infant is living or not [in the womb of the mother in labour], because according to the custom observed in many countries, if the child survives the mother, the father is the heir to all the movable effects; if, on the contrary, the child is dead before the mother, it is the mother's parents who inherit from her; so that if a court action arises between the father and the parents, as has often happened, it is for the surgeon to decide this; he is the authority who can win or lose the case for one or the other, and the judges will deliver their verdict only on his report; this is what he must engage to do with assurance from his conscience. Having performed the operation with all the precautions I have just indicated, if the infant is living, the kin shall have care of it; but if it is dead, it should be put back into the mother's womb, which should then be sewn up again in the same manner as one does with cadavers that have just been opened up." (p. 318).

phenomenon but, at the very most, a text that operated as a better established and more homogeneous framework. The resolution passed at the end of the discussion finally dismissed both types of registration without finding in favour of either, calling on principle for improvements but avoiding putting directors of offices of statistics in the difficult position of having to defend reforms that their respective governments would find too radical.[8]

4. THE EXPERIENCE OF LEGAL MEDICINE

Thus, in the mid-19th century, administrative statisticians seeking to establish homogeneous criteria free from ambiguity came up against the issue of infanticide, which was so closely interlinked with the issues of under-registration of births and declaration of stillbirths.

Research in the area of historical demography which sets out to reconstruct the population of 19th-century France mathematically, notably the work undertaken by Noël Bonneuil (1997), is a decisive element in assessing under-declaration. The process used by this demographer is based on statistical information specific to the female population, on which war has less impact than on the male population. From this base, Bonneuil took as his starting-point demographic information published throughout the century at the scale of *départements* by the French National Office of Statistics, at that time called the General Office of Statistics for France. He integrated this enormous whole into a single dynamic model that took account of elementary demographic dependencies between the different indicators (e.g., children born in one period must *in principle* be found among living persons, migrants or the dead in following periods). From this he inferred a coherent local, chronological reconstruction of the population. He then went back and compared this to the initial figures. The difference

[8] The following resolutions were adopted by the Congress in The Hague in 1869: "The governments of countries where declaration and registration of births are governed by the Napoleonic Code are invited to take measures that seem to them most appropriate to make known the number of children, 1. who enter the world already dead, and 2. are born alive but have died before the birth is declared. A child shall be considered stillborn when it has had at least six months of foetal life." (CIS, 1869, Vol. II, pp. 89–90); "and in other countries where the law recognizes true stillbirths, registration officials are enjoined to enter stillbirths into the registers as such, separate from deceased live births, at whatever time of life death took place, however short life has been"; "the Congress expresses the wish that in official returns of the movement of population, stillbirths should be categorized separately and should figure neither among births nor among deaths." (CIS, 1869, Vol. II, p. 92).

that he observed gives a homogeneous local, chronological indicator of under-registration of births. Bonneuil shows that, at the scale of France as a whole, under-registration would have declined regularly, from 14% or 15% at the beginning of the 19th century to under 6% at the end. His predecessor in this field, Etienne Van de Walle (1974), had himself reconstructed the female population of 19th-century France by another method, treating the *départements* less systematically. He, in fact, had found a rapid decline from 5.1% to 0%. Despite this difference in the size of decrease diagnosed, both methods highlight the same secular trend towards reduction in under-declaration. Bonneuil adds an assessment of the local variability of this trend, showing profiles of secular variations by *département.* From 1806 to 1836, the rates for his various categories of *départements* were clearly distinct: they were reported as 2.5%, 10%, 25%, 30%, 35% and 43% according to circumstances. By the end of the century, they were in the region of 5% or 10%. Thus arguably, as the century passed, there was a process of homogenization and reduction in under-registration. So, over that time, France went through a secular change in its statistical apparatus and the indices that this produced.[9] We can conclude not only that civil registration in France left a vague margin for declaration, but also that this margin diminished steadily throughout the 19th century.

The low level of registration of stillbirths, known to specialists at the time, did not prevent one of the founders of demography, Louis-Adolphe Bertillon – who was also a regular participant in the international congresses – from publishing a long review article on the topic, marking a milestone in the columns of Dechambre's *Dictionnaire encyclopédique des sciences médicales [Encyclopaedic dictionary of the medical sciences]* (1876). This comparative study of rates of stillbirth led to a conclusion destined for great success in demography: boys were said to be notably more frequent among stillbirths than among live births – an assertion to which we shall return later.

Finally, we have several studies on infanticide, undertaken by 19th-century medical jurists, with the most complete picture being given by Ambroise

[9] There is a literary example of a phantom birth being registered at a later date: the character Ferdinand du Tillet in Balzac's novel, *César Birotteau* (Paris, Gallimard, 1975, pp. 81–82). But, in Ferdinand's case, the incentive to regularize the situation comes from his professional ambition. Other evidence is provided by Ernest Legouvé, a man of letters in the same era (1849): "The Breton farmer whose wife brings a girl into the world still says today: my wife has had a miscarriage" or "Ask a peasant about his family and he will tell you: I have no children, just girls" (quoted in Armengaud, 1973; Vallin, 2002).

Tardieu (1818–1879), holder of the Chair of Legal Medicine in the Faculty of Medicine at Paris (1861) and a member of the Academy of Medicine (1858) (see Tardieu, 1868; see also Tourdes, 1889; Tourdes & Metzquer, 1896, which follow on from Tardieu). But the perspective of legal medicine is not that of administrative statistics. It does not aim to make the most complete assessment possible, but, in contrast, to provide expert intervention when the law is faced with a particularly doubtful case (see Box, below, on pp. 82–83).

Infanticide in France in the 19th century, according to Ambroise Tardieu, medical jurist

Cases of infanticide in France from 1851 to 1866

Cases brought	Number of accused	Men	Women	Acquitted
3012	3475	240	3235	1079
188 cases/year	100.0 %	6.9 %	93.1 %	31.1 %

Bodies of newborn infants deposited at the Paris Morgue from 1837 to 1866

Infants received	Coroner's post-mortem examinations	Confirmed infanticides
1244	1013	726
approx. 41 cases/year	100.0 %	71.7 %

Extracts from Tardieu's conclusions

It is in rural populations and among servants, as among domestic servants in the cities, that the greatest number of accused are to be found [...]. The dead bodies of newborn infants are found] sometimes [...] on the public road; sometimes the body is left on the threshold of a house, under a carriage entrance, in a driveway, very frequently inside a church. In the last case, it may happen that the infant was abandoned living in that place and perished as a consequence of abandonment; for it is not in remote spots that children whom it is wished to deliver alive to public charity are exposed. Also those who have suffered violent death and whom it is sought to hide away, are found especially in deserted places, on building sites, within an enclosure or a cemetery, where it is not rare for little bodies to have been thrown over the walls [...]. At other times, they have been cast into a sewer or a shaft. But perhaps the most frequent situation in the big cities is where they are taken out of a cesspool. It is a singular thing, but it seems that this is the most sure means of removing the traces of infanticide. The unhappy woman who has just given birth in secret and who has killed her infant, wastes no time in throwing her into the latrines and she believes herself assured of secrecy and of impunity [...]. (Tardieu clarifies that, although the movable cess tanks then in use in Paris were drained frequently, fixed cesspits were inspected only where there was suspicion).

Little bodies are also often thrown into a river, a brook, a pond, a pool, and in Paris into the canal. It should be noted in these cases that most usually these are not drowned infants, but newborns cast into the water after they have already been deprived of life. At other times, in the same conditions, they are found buried under the earth, at the bottom of a garden, in the corner of a wood or field, or even in a dunghill, where it is thought, not without reason, that the little bodies will be promptly consumed, or even in sand and in bags or barrels of bran and middlings. But it is still very common to find the body of the newborn in the mother's own bedroom, wrapped up, hidden in a cupboard, in a drawer in a piece of furniture, in a chest, a basket, a trunk, under a pile of rags, or under a bed and between mattresses. In the first moments that follow the crime, searches there most often yield results. Finally, so that nothing is omitted, cases must be cited where the remains of a body have been found in a fireplace, or in a stove, where someone has tried to get rid of it by burning it; and those where a body put behind a stove has become mummified there; or even those where fragments have remained at the bottom of a pot in which the body has actually been cooked; or in the pigs' trough, where it has been put for them to devour.

Means employed in the cases studied by Tardieu himself

Means employed	Cases	Proportions
Suffocation	281	50.6 %
Immersion in cesspool[10]	72	13.0 %
Fractured skull	70	12.6 %
Strangulation	60	10.8 %
Drowning	31	5.6 %
Neglect	14	2.5 %
Wounding	8	1.4 %
Burning	8	1.4 %
Umbilical haemorrhage	6	1.1 %
Exposure	3	0.5 %
Poisoning	2	0.4 %
Number of assessments between 1844 and 1868	555	approx. 22/year

Source: Tardieu (1868), pp. 8, 9, 11, 13–15, 99, 266–337.

> "There are no questions more arduous or more complex than those connected with proving the crime of infanticide scientifically [...]. However, I can claim to be the first scholar to make some personal observations on this question of infanticide [...]. I want to talk about the signs of death by suffocation, which [...] enable us to recognize and describe the obvious characteristics of the most usual method of infanticide, which, at the same time, is the one that, for lack of proof and sufficient evidence, most often escapes investigation." (Tardieu, 1868, pp. V–VIII).

Accounts from the administration of justice are well known to historians (Perrot & Robert, 1989). Various studies have been conducted by jurists, doctors or historians, scanning the last three centuries (Allexandre-Lefevre, 2002; Gaillard, 2000; Léauté, 1968; Tillier, 2001). As a result, infanticide as described by doctors and jurists partially conceals a particularly sombre aspect of the sexuality of those women least equipped to prevent an unforeseen pregnancy. Infanticide of newborn babies has been treated with varying degrees of severity from one era to another, but always in vain. In fact, its decline through the centuries has resulted from the interaction of a combination of new conditions: the elimination of female illiteracy, the near-disappearance of the employment of female servants and day labourers, improvements in legal medicine, the medicalization of pregnancy and, in the last decades of the 20th century, the rapid development of simple, effective methods of contraception. However, it is very rare for facts to be supplied in these studies that would allow us to know the sex of the victims. Thus Tardieu, who studied 555 cases of infanticide in the course of his career, supplied 59 reports in an appendix to his book published in 1868. Thirty-two of these reports indicate the sex of the child found dead: the majority – 21 of them – were girls. Beyond these few clues, it still remains for a historical survey to be undertaken on the topic. Few documents appear to provide material for such a survey; be that as it may, it would take us away from our central discussion here. However, we can state, following several of these authors, that, in France, infanticide is the one of the rare crimes that has almost disappeared as social density has increased (Gaillard, 2000; Léauté, 1968). Thus, the efforts of Tardieu and his successors in trying to establish probative, reproducible criteria for the qualification of infanticide were elements in a long-term process, in the course of which moral norms resulting from general legislation were extended even into the darkest areas of social life.

Medical jurists, specialists in public health, midwives, *accoucheurs* and obstetricians, statisticians and demographers – in the 19th and early 20th centuries, many such experts wanted to use their writings or their actions to create better conditions for pregnancy and delivery, and to improve registration of births. They attached no particular importance to the sex of the newborn child but, in wealthy countries, their actions contributed to an improvement and a relative social homogenization in the health and institutional conditions of birth, up to and including the stage of declaration before a registration official. It remains to be evaluated whether these general changes had any impact on the level of the sex ratio at birth or benefited both sexes equally – a point to which we shall return.

Chapter 4

SELECTION, SEXES AND STATISTICS (SINCE 1871)

1. SUCH AN INTRICATE PROBLEM

Thus, in accordance with the wishes of early-19th-century mathematicians, but at the cost of certain arbitrary decisions and empirical concessions that are now buried in the historical construction of counting procedures, the formation of specialized administrative offices in the 19th century provided a kind of laboratory at the scale of humanity, where figures on the proportion of the sexes of human beings at birth – and many other objects of statistical calculations – could be established. This quasi-laboratory has been functioning for two centuries now; despite the fact that it is highly imperfect, we have no more systematic device. The result is that, nowadays, it is most often possible for people to believe that there is no more banal or simple figure than the number of boys and girls born in the last year in a country meeting all the criteria for modernity. The other result is that remote episodes in the formation of this laboratory are simply not mentioned in recent works reviewing the analysis of sex ratios, especially from the point of view of the contemporary discipline of biology (Ericsson & Ericsson, 1999; Hardy, 2002; Majerus, 2003; or even Sober, 2005 – although he devotes some attention to a number of aspects neglected by his predecessors). Yet Charles Darwin (1809–1882) himself was well aware of these difficulties when he took the view that humankind alone – let us add, with its vast statistical apparatus – would provide him with satisfactory empirical material: "as the proportions [of the sexes] are known with certainty only in mankind, I will first give them as a standard of comparison" (1871, Vol. 1, p. 300). From this point of view, Darwin, although he hardly discussed Adolphe Quetelet's works, was on the same footing in some para-Quetelesian era: he shared the presuppositions resulting from the successful organization of statistics in Europe in his day, and took as "certain" figures that neither scholars of earlier generations nor even those who constructed statistical offices from

E. Brian and M. Jaisson, *The Descent of Human Sex Ratio at Birth*, 87–120.

the arcana of that era's international congresses would have so easily taken at face value.

For today's reader, Darwin's work forms the starting-point of modern biology. At the same time, among social scientists, it marks the beginnings of a broadly discredited academic movement that would lead to eugenics and to the simplisms of the various tendencies within socio-biology. Over the last decades, as in the case of the history of statistics, research within the history of sciences – with recent assessments given by Browne (1995, 2002) and Tort (1996) – has offered a more dense and subtle picture than the one preserved piecemeal in the collective memories of scholars. For our part, far from seeking to give a history, however brief, of biology or of Darwinian theory (on this subject, see Gayon, 1992, 1996; Moore & Desmond, 2004), or even a history of the latter's relationship with statistical calculation, we shall content ourselves with following the index of the human sex ratio in the work of Darwin and those who came after him. In this regard, does *The Descent of Man* (1871) mark a starting-point with no attachment to the elements of the intellectual history of the index, which we have already analysed? Since the appearance of this work and with the widespread use of later statistical techniques into biology, have mathematicians and social scientists paid any attention to the fortunes of human sex ratio at birth in the sphere of biology?

There is one feature that characterizes Darwin's research and the research that came in his wake, as compared to research in mathematics and the social sciences: that it accords with the resolution adopted in 1839 by the British Association for the Advancement of Science by taking humankind as an object of natural history (Browne, 1995, p. 421; [BA], 1840). Condorcet would have expected nothing less of the sciences during the utopian Tenth Epoch that he had assigned to the human mind (Condorcet, 1795 [2004]). But it will be remembered that the slightest trace of a "meliorist" perspective in his work had aroused a reaction from Malthus who, encountering such happy speculation on the future, challenged it on the grounds of inconsistency between the geometric progression of the population and the arithmetic progression of the means of subsistence (Malthus, 1798). Yet it was precisely by extending Reverend Malthus' assertion that Darwin defined the concept of natural selection in *The Origin of Species* (Darwin, 1859; Tort, 1996).

> "In the next chapter the Struggle for Existence amongst all organic beings throughout the world, which inevitably follows from |the high geometrical ratio of their increase, will be considered|. This is the doctrine

> of Malthus, applied to the whole animal and vegetable kingdoms. As many more individuals of each species are born than can possibly survive; and as, consequently, there is a frequently recurrent struggle for existence, it follows that any being, if it vary however slightly in any manner profitable to itself, under the complex and sometimes varying conditions of life, will have a better chance of surviving, and thus be naturally selected. From the strong principle of inheritance, any selected variety will tend to propagate its new and modified form." (1872, p. 3 [the marked passage between vertical bars thus, |, clarifies the initial version: "their high geometrical powers of increase, will be treated of", cf. 1859, pp. 4–5]; see also the section "Geometrical ratio of increase" in Chapter III of the same work).

Thus, the Darwinian empirical landscape was formed in three historical times and by a trick of reason. Firstly, in Paris at the end of the 18th century, in the context where metaphysics was challenging Süssmilch's style of natural theology and in the face of the excessively general nature of conclusions being drawn from the young Laplace's analytical calculus of probabilities, Condorcet indicated the moral conditions for a general physics of humankind and of animals, in order to illustrate his conception of asymptotic forms of progress of the human mind: the subject of history – the human mind – thus extended its horizons. Secondly, in London around 1800, the reception of the *Esquisse d'un tableau des progrès de l'Esprit humain* aroused in Malthus a neo-providentialist reaction which consisted of denying the human mind this status as the subject of history, putting forward instead the set of problems surrounding population and means of subsistence that arose from 18th-century political economy and were, in this context, peculiar to humankind. More than half a century later, in accordance with a resolution of the scholarly British Association for the Advancement of Science, which consolidated the empirical horizon expected by Condorcet, Darwin elaborated the concept of natural selection by extending to all species the contradiction that Malthus had raised precisely in opposition to the philosophy of history that the Association's action presupposed. At this secular scale, Darwin's founding concept appears to be a hybrid of two precedents that, earlier, had been in contradiction. In formulating it, Darwin was making an epistemological rupture that separated biological avatars for the study of human sex ratio at birth from their counterparts in demography (the principles of population according to Malthus, where the sex ratio really has no importance), sociology (this time, the consistency of the Durkheimian social fact; see Chapter 5) and mathematics (here, as we have seen, they were the various formulations of laws of large numbers). The animal nature of humankind, which was the very issue at stake in Darwin's gesture, was indeed to be neglected in demography until, one could say, the discipline accepted the predator-prey model formulated by Alfred Lotka (1880–1949) and Vito Volterra (1860–1940). The animal nature of humans has been excluded by sociology and by Durkheimian anthropology[1] – while mathematicians from Laplace onwards have been indifferent towards it.[2]

Having highlighted this divergence, we are now going to follow the "twig" of the biology of human sex ratio at birth. Such a choice does not mean that we are underestimating either former or current tensions among the sciences, such as those we have already noted among mathematicians, statisticians and physiologists – the "conflict of the faculties" that has been analysed by the sociology of sciences (Bourdieu, 1984, 2001). The tree of knowledge, with its branches of the different disciplines, is indeed no less perilous than the tree of life envisaged by Darwin himself.

[1] At this point, we shall merely mention Durkheim's critique of social theories inspired by a simplistic individualism that would come to be placed under Darwin's authority: "[...] we see from what precedes how false is the theory which makes egotism the point of departure for humanity, and altruism only a recent conquest. What gives this hypothesis authority in the eyes of certain persons is that it appears to be the logical consequence of the principles of Darwinism. In the name of the dogma of struggle for existence and natural selection, they paint for us in the saddest colours this primitive humanity whose hunger and thirst, always badly satisfied, were their only passions; those sombre times when men had no other care and no other occupation than to quarrel with one another over their miserable nourishment. To react against those retrospective reveries of the philosophy of the eighteenth century and also against certain religious doctrines, to show with some force that the paradise lost is not behind us and that there is in our past nothing to regret, they believe we ought to make it dreary and belittle it systematically. Nothing is less scientific than this prejudice in the opposite direction. If the hypotheses of Darwin have a moral use, it is with more reserve and measure than in other sciences. They overlook the essential element of moral life, that is, the moderating influence that society exercises over its members, which tempers and neutralises the brutal action of the struggle for existence and selection. Wherever there are societies, there is altruism, because there is solidarity." (Durkheim, 1893, [1978, pp. 173–174]; translation from Simpson in Durkheim, 1933, pp. 196–197). See the following chapter for Maurice Halbwachs' development of this critique.

[2] As we can see, each discipline – biology, sociology, demography and the various forms of mathematics, each according to its own temporality – appears in this instance to be enduringly characterized by a particular historically constituted relationship between learned subject and scientific object. On similar issues, see Panofsky (1927b) and Brian (1994a). In particular, the formation of the concept of *natural selection* consolidated and effected a change in this relationship, a change which is specific to biology.

This pastiche of an extract from *The Origin of Species* 1872, p. 104, in which we substitute "disciplines" for "species", and "science" for "life", is pertinent: *the green and budding twigs may represent existing* disciplines; *and those produced during former years may represent the long succession of extinct* disciplines. *At each period of growth all the growing twigs have tried to branch out on all sides, and to overtop and kill the surrounding twigs and branches, in the same manner as* disciplines *and groups of* disciplines *have at all times overmastered other* disciplines *in the great battle for* science. *Of the many twigs which flourished when the tree was a mere bush, only two or three, now grown into great branches, yet survive and bear the other branches.*

But historians of ideas have very often tended to clip back this lush landscape of past works to create formal gardens... Thus, nowadays, it is the name of Ronald A. Fisher (1890–1962) that marks the collective memory of biologists in relation to the sex ratio. His 1930 argument on species whose reproduction proceeds from the differentiation of two sexes – a principle rather than an empirical conclusion, but in his eyes a kind of theorem – is that from the point of view of natural selection there is no trend towards an advantage for one of the two sexes to predominate numerically over the other at birth. The sex ratio should therefore tend towards 50% for most species (Fisher, 1930; Hamilton, 1967; Majerus, 2003; Williams, 1975). The robustness of Fisher's argument led biologists to reread some of their predecessors, especially Darwin, with a new eye. So for about 20 years now, it has been accepted that the first edition of *The Descent of Man* (1871) included a sort of anticipation of Fisher's argument, which would be erased from the second edition three years later (Edwards, 1998; Sober, 1984), and that some ten years later again, Carl Düsing, the Jena physiologist born in 1859, would arrive in his own way at a comparable formulation (Edwards, 2000). But reading these authors suggests that the path from *The Descent of Man* to Fisher's principle is even more overgrown, and that it can be further illuminated by a comparison with the works of Francis Galton (1822–1911) and the thesis of the Italian bio-statistician Corrado Gini (1884–1965), who is no less important in the history of statistics. Consequently some particular aspects of the use of mathematical calculations in biology require clarification here.

2. DARWIN'S ARGUMENTS

The two main editions of *The Descent of Man*, from 1871 and 1874, present a collection of significant variations (see the critical apparatus of Appendix B). We know the reasons for this: Darwin devoted a great deal of attention to responding to a number of objections that the publication of his works had encountered and to integrating new elements in support of his theses (Browne, 2002; Introduction by Tort in Darwin, 2000). We also know that Darwin mobilized considerable empirical material through the most varied channels: personal observations; long, in-depth exchanges with favoured scholars; information sent now and again by correspondents; or even occasional gleanings and opportune interactions – as when, for example, between the first two editions of *The Descent of Man*, his Dutch translator, Hermanus Hartogh Heys van Zouteveen, obtained some figures for him on the births of white people and slaves at the Cape of Good Hope; these had previously been published by Quetelet, albeit with some reservations, and editors of *Physique sociale* have still for some years been hard pressed to identify their source (Quetelet, 1997, pp. 60–61). Like a number of contemporary scholars, Darwin mixed figures with no critical concern or, at any rate, no attention to their origin.

What material on the proportion of the sexes at birth did Darwin have at his disposal? The comparison with two of his elders, Quetelet and Louis-René Villermé (1782–1863), the French statistician and doctor in the field of public health, is illuminating: both of them commented on the same works published in German, English or French (Quetelet, 1835, 1869; Villermé, 1832a, 1832b, 1832c). Laplace, Poisson, Hofacker (that is, Notter) and Girou de Buzareingues, for example, appear everywhere. Only the channels for their information differ, and consequently the place that comments on these authors occupy in the general arrangement of the works of Quetelet or Villermé also differs. In this regard, one particular work seems to have guided Darwin's approach to the distribution of the sexes at birth. This was a paper that had appeared more than 30 years earlier, addressed by Charles Babbage (1791–1871) to the editor of the *Edinburgh Journal of Science* (Babbage, 1829). We know that Babbage was one of Darwin's principal interlocutors on the subject of statistical questions of births and on the works of Quetelet (Browne, 1995, p. 385).

The mathematician Babbage, who today is most often remembered for his contribution to the history of the mechanization of calculation, was an eminent figure in Darwin's circle of scientific and intellectual patrons (Browne, 1995). It was the Analytical Society, founded at Cambridge in 1812, that ensured the promotion of the analytical conception of mathematics current in Continental Europe, at a time when the synthetic model of Newtonian mathematical sciences was predominant in Great Britain. As members of this circle, Babbage and those close to him – John

F. W. Herschel (1792–1871), George Peacock (1791–1858), then later Augustus De Morgan (1806–1871) and George Boole (1815–1864) – were most attentive to all late-18th- and early-19th-century French innovations in the mathematical sciences. Reforming English science by means of Continental European science, in the military and political context of this period, did not just mean adopting results and methods found elsewhere: it also meant bringing English science out of its inturned position – a result of the period of the Napoleonic blockade – and deliberately seeking in France and Germany everything that might help to strengthen the assertion of a young generation of scientists. The Society's most obvious contribution was the translation of one of the best textbooks on analysis written in French in this era, by Silvestre Lacroix (1765–1843), who for a long time assisted Condorcet in his teaching (Lacroix, 1802, 1816): this promoted Leibnizian notations of the integral calculus, which were certainly more powerful, to the detriment of the Newtonian notations still in use in England (Babbage, 1864; Durand-Richard, 1996, 2001; Enros, 1979).

The 1829 article arose from tension between Babbage and De Morgan over what mathematicians should say about the business of insurance. Condorcet and Laplace had themselves touched on this question in Paris 50 years earlier. Both of them, while they were working on developing the analytical calculus of probabilities, also from time to time carried out assessment duties at the Royal Academy of Sciences of Paris, since, towards the end of the *Ancien régime*, this learned Company was having to express an opinion on the validity of various insurance companies' plans (Baker, 1975; Brian, 1994a; Condorcet, 1994). At half a century's remove, there were some equivalent circumstances in both France and Britain. However, they were inscribed in different institutional frameworks and social divisions of the tasks of monitoring the insurance trade (Campbell-Kelly, 1994). De Morgan had criticized the collection that Babbage, then in the service of the Protector Life Assurance Society, devoted to the comparison of different life insurance formulas (Babbage, 1826; see also De Morgan, 1838). The article that appeared in Edinburgh in 1829 was presented in the form of a letter to the editor of the *Journal*, accompanied by records of statistical tables collected by the author from some well-informed correspondents. In 1826, Babbage had been an advocate of the development of the insurance industry. On this subject, he had pleaded the cause for trust in the application of the calculus of probabilities. Three years later, he was commenting ironically on reservations that De Morgan had expressed in the meantime. After his critique had recalled the variability of the observations published by the Office of Longitudes in Paris, Babbage set out to show that he was not ignorant of variations observed in the proportion of the two sexes at birth, or of the importance of these for the insurance trade, since "facts and accurate enumerations are the great and only bases on which such transactions can securely rest" (Babbage, 1829, p. 85). In doing so, Babbage recalled, citing Laplace, that the proportion of the sexes among children received at the Paris Hospital was exceptional. In these conditions, he discussed – with supporting figures – three linked scenarios: illegitimate births, differences between city and countryside, and stillborn infants. He then added some elements to the picture of the objection that Laplace had raised himself almost 20 years earlier, and which Fourier had further elaborated by wishing for still more organized observations. The phenomenon, Babbage acknowledged, presented indices of variability. He implied that there were no reasons for doubt, but many empirical investigations to be undertaken. Babbage next drew the attention of scholars to the tables that three of his informants had obtained for him since 1826; one was the compiler of statistics for the Dublin Foundling Hospital, another the "*Chef de Division et Directeur de Bureau Statistique dans la ministére de l'Interieure*" (sic) for Westphalia and the third, Johann Gottfried Hoffmann (1765–1847), head of the Office of Statistics in Berlin. Here Babbage added a mention of research presented in 1827 to the Paris Academy of Sciences, but without indicating its author. These German and French elements came from his recent travels in Europe.

Johann Gottfried Hoffmann was very preoccupied with the statistics of religion in Prussia and especially with measuring what he described, at the end of his life, as "*Judenfrage*", thus creating the empirical material for a way of thinking about the place of Jews in Prussian society, which was to assume a terrible scale (Hoffmann, 1842, 1844). He was an economist of the first importance in his day (Schuster, 1908). His statistical publications contributed strongly to legitimizing anti-semitism in well-read circles in Germany (Keval, 1999). His articles in the 1840s show that, from the 1820s onwards, he was establishing statistics on Jews in Prussia. He communicated the preliminary elements of this work to Babbage, who published them as an appendix to his letter to the *Edinburgh Journal of Science*. Babbage introduced them in the following terms, after discussing variations in the proportion of births according to sex had already been shown before him:

> "I shall notice one other circumstance connected with this subject. It is the remarkable excess of males amongst the children of the Jews in

Prussia. For every ten thousand females born amongst them there are 11,292 males [53.03%]." (Babbage, 1829, p. 91)[3]

What is implied by this mention of "one other circumstance"? The context here remains that of religion, and is not yet completely racial in the sense of a naturalist's conception of human beings. Because, on Babbage's part, the background to the discussion of the proportion of the sexes at birth remained a critique of the physico-theology so powerful in England in the 18th century, it was in fact the matter of polygamy that was the issue here. This is why Babbage immediately continued his argument by merging the Jewish and Muslim religions:

> "It would be interesting to examine this fact amongst the Jews in other countries, and still more so, could we procure any correct enumeration of births in any country in which the Mahometan religion prevails." (Babbage, 1829, p. 91)

Darwin, in 1871, although he was to rely heavily on Babbage's text, would not retain this wording – but he did come back to it in another way. In 1829, Babbage next went on to make another curious observation, moving seamlessly from religions compatible with polygamy to domestic animals:

> "I cannot conclude this subject, without recalling to your notice a statement, in the History of the Academy of Sciences of Paris [*sic*] for the year 1827. It is stated as the result of some experiments lately tried, that

[3] Babbage appended the tables that Hoffman had obtained for him. One of them compares marriages, births and deaths of "Christians" and "Jews" in six Prussian provinces during the years 1820–1824 (Babbage, 1829, Table XI, p. 102). Here are the birth figures in summary:

	Boys (M)	Females (F)	Births (N)	Sex ratio (M/N)	CI95% $(1/\sqrt{N})$	Minimum	Maximum
Christians	842 894	794 580	1 637 474	51.48%	0.08%	51.40%	51.55%
Jews	12 454	11 029	23 483	53.03%	0.65%	52.38%	53.69%

At a 95% level, the two confidence intervals do not overlap. The difference between the proportions, insofar as counting was consistent, is relevant. But conditions for civil registration of births in the early 19th century were not consistent for all religious faiths, any more than was the statistical coverage of populations in that era. Nor do we have any information here on the distribution of births between cities and the countryside, a criterion then accepted as relevant. Yet the figures for the six provinces totalled above suggest strong regional differences. The only thing being measured here is the overall difference in the proportions of the sexes in the official entries of infants in Prussian registers (on the problems posed by registration statistics in 19th-century Europe, see the preceding chapter).

in a flock of sheep consisting of 71 females and 61 males, by selecting strong females and young males, and by feeding the females high and not the males, the result was amongst the births

	Males	Females	
	53	84	
by the reverse process	80	50"	(Babbage, 1829, p. 91)

In 1829, people would hardly have known what Babbage was talking about! The report on activities for the year 1827, traditionally known as the *Histoire de l'Académie des sciences* in the circles of the Paris Company, edited by Georges Cuvier, Permanent Secretary for the division of Physical Sciences (in the sense of natural sciences), would not appear until 1831. Therefore, in 1829, Babbage was either referring to a meeting that he had been told about during his stay in Paris a year earlier, or else was going back to notes obtained for him, taken at the reading of a paper. (To our knowledge, there is nothing to indicate that he attended meetings at the Paris Academy of Sciences.) In short, he made a show of competence in response to De Morgan, although in fact his account was somewhat inaccurate, in that the figures differ from the report that appeared two years later; moreover, the case of "71 females and 61 males" was a third experiment, and would nowadays be described as a control group. But that is not all: both the academic volume for 1831 and the minutes of the Paris meetings for the year 1827 prove unambiguously that the experiments mentioned by Babbage are those of Girou de Buzareingues (see Chapter 2).

> "Some curious experiments, not only for agriculture, but for general physiology, are those of Monsieur Girou de Busareingues [*sic*] on the procreation of the sexes. It is on the greater or lesser comparative vigour of the individuals which are mated that the sex of the product depends. If it is desired to have more females, young males and females in the prime of life should be used, and the females should be fed more abundantly than the males. If it is desired to produce more males, the opposite must be done. With the first process, 84 females as against 53 males were obtained from one lambing, and with the second, 55 ewes as against 80 males; while an equality of strength and feeding in the same flock gave 71 females and 61 males. Birds follow the same law as sheep. In the same farmyard, the strongest females give a greater number of individuals of their sex than small females; young females who have not reached full development give more males." (Cuvier, 1831, p. CLXXXVII)

As for Darwin, Babbage's 1829 article offers evidence of the bases for discussion between the mathematician and the young naturalist about the distribution of births by sex. We can easily find many clues to this in *The Descent of Man* by highlighting the references made to authors and issues mentioned by Babbage in his article. We should note that this discussion fits with the resolution passed by the British Association in 1839 on the natural character of humankind, and that it presupposes a particular syntax of known things: the proportion of the sexes at birth offers a very general constancy at the scale of humankind; the question of empirical variability within the human species remains to be dealt with; the variations established generally relate to natural births and stillborn infants; a continuum of cases would become apparent, running in order from monogamous peoples to various species of animals, through polygamous peoples and domestics species.

The supplement to Chapter VIII of *The Descent of Man* deals explicitly with the sex ratios of human and animal species. It is a sort of updating of Babbage's 1829 outline assessment, which Darwin explicitly gives as an excursus, once he has set out the concepts of natural selection and sexual selection. In order to understand how the human sex ratio at birth plays a part in Darwin's thought, it is useful to situate this particular element within the general framework of his theory. Darwin himself clearly indicated the object of *The Descent of Man* in his Introduction (Browne, 2002).

> "During many years I collected notes on the origin or descent of man, without any intention of publishing on the subject, but rather with the determination not to publish, as I thought that I should thus only add to the prejudices against my views. It seemed to me sufficient to indicate, in the first edition of my *Origin of Species* [1859], that by this work 'light would be thrown on the origin of man and his history'; and this implies that man must be included with other organic beings in any general conclusion respecting his manner of appearance on this earth. Now the case wears a wholly different aspect. When a naturalist like Carl Vogt ventures to say in his address as President of the National Institution of Geneva (1869), *'personne, en Europe au moins, n'ose plus soutenir la création indépendante et de toutes pièces, des espèces'*, it is manifest that at least a large number of naturalists must admit that species are the modified descendants of other species; and this especially holds good with the younger and rising naturalists. [...]
>
> In consequence of the views now adopted by most naturalists, and which will ultimately, as in every other case, be followed by others who

are not scientific, I have been led to put together my notes, so as to see how far the general conclusions arrived at in my former works were applicable to man. [...]" (Darwin, 1871, Vol. 1, pp. 1–2).

In this second major work, as in his preceding one, Darwin did not envisage the diversity of human phenomena (or of those specific to other species) either in the manner of a mathematics of the probable as in the work of Condorcet or Laplace, who had seen diversity as a cause for uncertainty, or in the manner of Quetelet's social physics, based on an analogy between the variations of observed cases and those of measurement errors. For Darwin, the variability of individuals within each species is the precise scene of the action of the founding concept of the theory: natural selection.

"It may be said that natural selection is daily and hourly scrutinising, throughout the world, every variation, even the slightest; rejecting that which is bad, preserving and adding up all that is good; silently and insensibly working, whenever and wherever opportunity offers, at the improvement of each organic being in relation to its organic and inorganic conditions of life. We see nothing of these slow changes in progress, until the hand of time has marked the long lapse of ages, and then so imperfect is our view into long past geological ages, that we only see that the forms of life are now different from what they formerly were" (Darwin, 1859, p. 84).

So there is no sense in seeking to measure – like Laplace – a general cause or – like Quetelet – an average man. Similarly, to hope for an immediate grasp of the object through statistical surveys is completely futile: the object is revealed by the comparison of one of species to another (whether disappeared or living), or even by other more subtle comparisons, and not by a hypothetical balance of "the strong carrying the weak", as used to be said (Brian, 1991; Perrot, 1992; and, for an early genesis of this expression, Hamon, 2001). There is nothing astonishing, therefore, in the fact that Darwin so often took a systematically opposed view to Quetelet's empirical arguments, even though reference to them in his work was most often implied.

"It is manifest that man is now subject to much variability. No two individuals of the same race are quite alike. We may compare millions of faces, and each will be distinct. There is an equally great amount of diversity in the proportions and dimensions of the various parts of the body; the length of the legs being one of the most variable points. [...] The variability or diversity of the mental faculties in men of the same race, not to mention the greater differences between the men of distinct races, is so notorious that not a word need here be said. So it is with the lower animals. [...] With respect to the causes of variability, we are in all cases very ignorant; but we can see that in man as in the lower animals, they stand in some relation to the conditions to which each species has been exposed, during several generations." (Darwin, 1874, pp. 26–27)

But, in these conditions, where then is the consistency of the object which guarantees that it can be observed over a long time scale? The Darwinian conception of variability was questioned and the possibility raised by his contemporaries – notably the engineer Fleeming Jenkin (1833–1885) in 1867 – of putting it to the empirical proof. But from Darwin's point of view, such criticisms were unfounded. This was because, in his eyes, the consistency of the object in his concept lies in the transmission of characteristics and therefore in the fact of their heredity.

"I have elsewhere [(Darwin, 1868)] so fully discussed the subject of Inheritance, that I need here add hardly anything. A greater number of facts have been collected with respect to the transmission of the most trifling, as well as of the most important characters in man, than in any of the lower animals; though the facts are copious enough with respect to the latter." (Darwin, 1874, p. 27)

The next stage of our reconstruction looks at the place that Darwin gave to the proportion of births in the development of successive editions of *The Descent of Man.* He indicated in his introduction the way the first part of the work would be arranged. "The sole object of this work is to consider, firstly, whether man, like every other species, is descended from some pre-existing form ...": this is the first chapter; "... secondly, the manner of his development ...": these are Chapters II to VI; "... and thirdly, the value of the differences between the so-called races of man": this is Chapter VII. But the work includes no less than 14 other chapters, grouped into two further parts; several commentators have seen this as a second essay, something that Darwin himself acknowledged.

"During many years it has seemed to me highly probable that sexual selection has played an important part in differentiating the races of man; but in my *Origin of Species* I contented myself by merely alluding to this belief. When I came to apply this view to man, I found it indispensable to treat the whole subject in full detail. Consequently the second part of the present work, treating of sexual selection, has extended to an inordinate length, compared with the first part; but this could not be avoided." (Darwin, 1874, p. 3).

For Darwin, the distinction of the races was an empirical starting-point whose causes remained to be explained. Let us clarify, in order to prevent any remaining ambiguity, that here we do not intend to confirm such a position, nor even to enter into debate about its interpretation (on this subject, see Tort's Introduction to the French translation of Darwin, published in 2000; and contributions to Tort, 1996). We are seeking to recreate the logic of his argument in order to understand the place that the distribution of the sexes at birth might occupy in it. Darwin's search for the causes of the human races led him to sexual selection, and sexual selection, as we shall see, to the numerical distribution of the sexes.

> "We have thus far been baffled in all our attempts to account for the differences between the races of man; but there remains one important agency, namely Sexual Selection, which appears to have acted powerfully on man, as on many other animals. I do not intend to assert that sexual selection will account for all the differences between the races. [...] Nor do I pretend that the effects of sexual selection can be indicated with scientific precision; but it can be shewn that it would be an inexplicable fact if man had not been modified by this agency, which appears to have acted powerfully on innumerable animals. It can further be shewn that the differences between the races of man, as in colour, hairiness, form of features, &c., are of a kind which might have been expected to come under the influence of sexual selection. But in order to treat this subject properly, I have found it necessary to pass the whole animal kingdom in review. I have therefore devoted to it the Second Part of this work." (Darwin, 1874, pp. 198–199).

Darwin defined the principles of sexual selection in the next chapter – the eighth chapter of the book, and the first in the second part – ("[Sexual selection] depends on the advantage which certain individuals have over others of the same sex and species solely in respect of reproduction", 1874, p. 209). There is a differential advantage with regard to reproduction among individuals of the same sex within the same species. As this reproduction results from the encounter of the two sexes in the species, their respective proportions are important. Thus, the numerical proportion of the two sexes is a major requirement for the possibility of sexual selection in a given species (Extract 3, Appendix B). Darwin therefore had to take stock of what we call the sex ratio. This was the purpose of the "*Supplement on the proportional numbers of the two sexes in animals belonging to various classes*" (1864, pp. 242–260), appended to Chapter VIII, where he gave his dossier of observations with commentaries. The two versions of this

Supplement, from 1871 and 1874, had certain differences that have attracted the attention of historians.

Let us start by considering the first edition (1871). In the section of Chapter VIII that deals with the numerical proportion of the two sexes (and which includes a cross-reference to the detailed discussion reserved for the Supplement), Darwin noted that there were scant empirical sources, although there were perhaps a few known elements regarding domestic animals. The result was that he would acknowledge, in the Supplement, that the human species alone offered empirical material that he considered to be a "standard of comparison" (1871, Vol. 1, p. 300; Extract 1, Appendix B).[4] But the fact of the matter was that, from his starting-point, Darwin took the view that, for the domestic animals that constitute a quasi-experimental case of sexual selection, the proportion of each sex at birth would be close to 50%: "By indirect means, however, I have collected a considerable body of statistical data, from which it appears that with most of our domestic animals the sexes are nearly equal at birth" (1871, Vol. 1, p. 263; *ibid.*). This was a first argument, purely empirical, formulated by Darwin about the sex ratio at birth. But it was a lesser argument for Darwin, because, in his eyes, what we nowadays call "secondary sex ratio" was only second-best. What was important in sexual selection was "tertiary sex ratio" – that is, sex ratio at the moment of reproduction. But Darwin's text revolves around this concept without really defining it precisely when he writes "proportional numbers of the two sexes" – for the only *enumerations* that he could mobilize were either of proportions at birth (the secondary sex ratio) or of overall proportions out of the total numbers of the two sexes in a given population. The result of this was that, on several occasions in these sections, he explored these two imperfect indices in order to give a verdict on the phenomenon that interested him. Darwin had at his disposal two precise concepts (natural selection and sexual selection), but inadequate indices. Hence the fact that a number of his explanations in *The Descent of Man* marked out the boundary of some later biological research, where

[4] This point in our historical survey is not the place to reproach Darwin for having shown an improper objectivism when he took the view that enumerations were external to the conditions peculiar to a given species. Indeed, it would not be until the end of the 20th century that such a critique could be formulated within the social sciences. It remains no less true that, from the anachronistic point of view that might be that of a scholar today, this criticism should be taken seriously: how can one conceive of the natural character of humankind without taking into account the fact that this species has developed means of counting that give it reflexive self-regulation criteria?

the analysis of stocks (the proportion of the two sexes in general) and the analysis of flows (their distribution at birth) can be found together even nowadays.

Since enumerations of the human species were the best ones, the Supplement to Chapter VIII gave the most salient results of these (Extract 2, Appendix B). As far as its syntax is concerned, Darwin's case file is not really different from Babbage's. But his sources were newer. This was because he was able to draw on publications of official British statistics (Registrar-General, 1869), although he did not know the comparative statistics on the same subject produced in the wake of the International Statistical Congress (Quetelet & Heuschling, 1865). Once he had indicated the national figures, Darwin gave a few recent gleanings as used by Babbage: these were first of all the comparison of Jews and Christians, then illegitimate births, then stillbirths.

For the first of these, he relied on a work by the Geneva botanist Marc Thury (1822–1905) (Extract 2, Appendix B). Nowadays, this publication seems rather astounding. It was a pamphlet intended to promote a theory of generation in the style of Girou (whose work, incidentally, is cited extensively), which includes a "Note IV", of which Paragraph 5 presents, in a few lines taken for granted by other authors, the figures that Darwin reproduced (Thury, 1863). Thury cites a classic work of physiology – the treatise by Carl Friedrich Burdach (1776–1847) published 30 years earlier in Leipzig (Burdach, 1835–1840), with the French translation following very shortly in 1837–1841. This translation and the erroneous figures that it included were Thury's source. He added the allusion: "the reader will perhaps find a likely explanation of these facts in the probability that Jews more fully observe certain prescriptions of the law of Moses" (p. 25): so – if we hold to his theory of generation – the explanation lies in the effect of these observances on rates of sexual relations between the parents, and even perhaps the effect of dietary restrictions. Yet Burdach, in *Die Physiologie als Erfahrungswissenschaft [Physiology as a science of experience]*, in the manner of the German Romantic science of his day (Poggi, 1994), summarized the things known before him about the proportion of the sexes at birth (Süssmilch, Spallanzani, Fourier, the Paris Office of Longitudes, Girou, "Hofacker", Sadler, Quetelet) and added to them a few new curiosities about Jews, described in terms of race, which were drawn from an article recently published in the *Zeitschrift für die Staatsarzneikunde* supplemented by notations obtained by a zealous collaborator... This was how word-of-mouth between scholars worked: at the margins of an almost unchanging corpus lay a small catalogue of curiosities, which were never very reliable; clues to these spread so obstinately – in the mid-19th century, they circulated by two international scientific routes, both of which started in Prussia and arrived in London half a century later – that trust in observations, however unverifiable, became stronger; and, in this specific case, a racial presupposition was reinforced.

Let us take up the thread of Darwin's arguments. Next comes the argument on the excess of girls among illegitimate births: "a still more singular fact", supported by a reference to Babbage's 1829 article, and presented as a general fact, almost in the terms that Laplace had reserved for the "greater facility in the birth of boys" (Laplace, 1825): "in different nations, under different conditions and climates, in Naples, Prussia, Westphalia, France and England" (1871, Vol. 1, p. 301). But Darwin then passes on to the case of stillborn infants, giving a few facts without interpreting them (1871, Vol. 1, p. 302). He brings his review to an end by indicating other factors discussed before him, and notably the possibility of a parental age gap effect.

Having reviewed the known figures for animal species, Darwin then sums up the *Supplement*. In doing so, he discusses the possibility, from the point of view of natural selection (not from the point of view of sexual selection), of an inequality in the proportions of the two sexes at birth. He reasons firstly by hypothesizing that this trait might be favourable to a species; but he observes that individuals do not derive any advantage or disadvantage from it. Therefore he concludes that it is difficult to accept that natural selection is at work. Then he takes the question almost *ad absurdum*, in a passage that is certainly fairly complex. He supposes an unknown cause which disturbs the equality between proportions of the two sexes, and he reasons combinatorially on the propensities of the generations to preserve this difference. He concludes that there is a trend towards re-establishing equality, following a structure of argument very close to that of Condorcet – whom he had undoubtedly not read – who had reasoned from the moral point of view. Darwin asserts that there is no reason to accept that natural selection may strengthen an imbalance between the sexes at birth. He then extrapolates this conclusion: "we may conclude that natural selection will always tend, though sometimes inefficiently, to equalise the relative numbers of the two sexes." (1871, Vol. 1, p. 318; Extract 7, Appendix B). It is this extrapolation, more than the line of reasoning actually followed by Darwin, which reminded people, after Fisher, of the principle linked with his name.

It is true that this passage, specific to the 1871 edition, is not very clear and that it includes a fairly obscure reference to the "doctrine of chances". Should we be surprised, therefore, that Darwin revised the *Supplement* for the second edition? Although several commentators have noted the sacrifice of this extrapolation (it might not be missed, but surely this is at the price of anachronism?), no attention has been paid until now to the fact that the empirical material given by Darwin in the 1874 edition no longer works in the same way. There is, of course, the addition of the proportions of births according to sex among masters and slaves at the Cape of Good Hope, derived from Quetelet's *Physique sociale* and obtained by a Dutch translator. However, this is a curious observation about the races, as flimsy as the earlier ones. But the important point for our purposes is that the 1874 additions transform Babbage's material into a proof of the capacity of sexual selection to partly explain disproportions in the secondary sex ratio in the case of illegitimate births and in the case of stillbirths (Extracts 4 and 5, Appendix B). These two partial conclusions come on top of the fact that Darwin found it impossible to demonstrate any effect of natural selection on the differences in the proportion of the sexes at birth. And as we have seen, this proportion at birth was only an indirect means of capturing its counterpart at the moment of procreation. Hence the conclusion he gave in 1874: "I now see that the whole problem is so intricate that it is safer to leave its solution for the future." (1874, p. 260). Therefore there is no abrupt reversal on Darwin's part in 1874, but an increased caution, once new elements have been accepted.

As we reach the end of this particular path, it must be said that the proportion of the sexes at birth plays a completely secondary role in Darwinian theory. The numerical ratios of the two sexes do matter, but only very much later, at the time of reproduction. Certainly, Darwin brought in everything that 19th-century statistics had produced before him by way of figures on this subject, thus sharing the numerical evidence of his contemporaries. But his construction avoided the most legitimate alternative before him, between the purely probabilistic approach and the exploration of social physics. The object of Darwinian biology is variability, and its scale is that of the formation of species. This leaves no room for any "law of large numbers", either in terms of the mathematics of the probable in the sense of Laplace or Poisson, or along the lines of Quetelet's physics of the average man. After Darwin, several statisticians, even Fisher, as we shall see, sought to revive earlier approaches, with some success.

3. WAYS THROUGH THE LABYRINTH

Now let us return to some of Babbage's and Darwin's data. Girou's experiments, like the extrapolation by his commentators in the 1825 *Annales des sciences naturelles* – as we have seen – and then Notter's experiments, as detailed in his thesis, were clearly produced using old forms of numerical methods in agronomy. Nowadays one might try to interpret them in terms of experimental plans, as defined from the point of view of mathematical statistics by Ronald A. Fisher almost a century later (Fisher, 1925, 1926, 1935; Fisher & Wishart, 1930). The coherence of the works of three 19th-century authors and their presence in French and German academies and universities show that there was indeed a method there – a technique, as Condorcet would have said – accepted as such by specialists. The analytical calculus of probabilities obviously had no place in it. The method originated in those in current use in early-19th-century agronomy. This was a second feature that had something in common with the development of experimental plans in Fisher's sense, a century later in Great Britain. But, before the international success of thinking based on the calculation of the average in the mid-19th century, Girou, his enthusiastic commentators and Notter offered an account – which one is tempted to describe as "pre-Quetelesian" – of this statistical technique. As we have seen, Quetelet and Babbage, although both mindful of the variability of phenomena – but both at first inclined to seek a central trend, because of their adoption of a post-Laplacian point of view – did not perceive the systematic combinatorial nature of the experimental constructions whose results they reported. And it is only once the conception of the average had entered the mind of statisticians – the sociologist Maurice Halbwachs, in about 1940, would have said the "collective memory of statisticians" – and regarded as obvious, that Francis Galton (1822–1911), then later Fisher and others were able to devise a statistics of variability understood first and foremost as that of differences from the average trend.[5]

The interpretation given by Stephen Stigler, the historian of sciences, of Galton's works as an "English breakthrough", is pertinent (Stigler, 1986, pp. 265–299). There was indeed an English style that was to spread beyond this initial base in successive waves punctuated by the two World Wars of the 20th century. It was on the outer edges of the world of European statistical offices in the second half of the 19th century that mathematical statistics

[5] Thus, Quetelet's average is not just a calculation method. Historically and epistemologically, it has operated as a symbolic form (Brian, 1991; Panofsky, 1927a).

developed in England as a result of Galton's works, becoming inseparable from eugenics (Beaud & Prévost, 2000; Schweber, 1997; Szreter, 1996). Among other starting-points were his book *Hereditary genius: an inquiry into its laws and consequences* (1869) and his article "Regression towards mediocrity in hereditary stature" (1886). Retrospectively, this way of seeing statistics seems to us to be a breakthrough because, with it, variability around central trends was constituted as the object of the method. It subverted, by combining its elements differently, the particular relationship that Quetelet had wanted to establish between the organization of the construction of empirical counting and the demonstration, at the scale of humanity, of the average type. But with post-Galtonian statistics, it was also the field of application of the methods that was transformed by the time of the Second World War: the statistics of variability was first and foremost a statistics of biological phenomena (Stigler, 1986, 1999). We know that, although Darwin integrated results obtained by his younger cousin Galton into his arguments, the younger man did not immediately adopt the biological point of view of his elder (see, in particular, Browne, 1995). The fact that he was won over and the real affinity between their empirical concepts – the same tracking of the variability of things, encouraged in the early-19th-century English intellectual and institutional context – were to seal the fate of this English mathematical statistics. Yet, in Galton's work, there still remained the "end of the Enlightenment" problems of the transmission of qualities, while the exploration of the proportion of births according to sex was left in abeyance. As we have seen, there was nothing in Darwin's work to give Galton an incentive to return to it. Against the background of natural theology, present in the context in which these two authors worked, to leave the general cause of the excess of male births unresolved was indeed their *hypotheses non fingo* (Newton, 1713).

Such caution did not cross the Channel. In Germany and in Italy, works on the proportion of the sexes at birth recorded the results of the rise in the production of statistics and in statistical thinking throughout the 19th century. The doubts and thoughts of the German professor of statistics Wilhelm Lexis (1837–1914) about Poisson's conception of the binomial dispersion (1830) in effect prepared the ground for a particular "physico-mathematical" route (Lexis, 1876; discussed by Stigler, 1986, pp. 222–225). One of his students at the *Universität Straßburg* (that is, the University of Strasbourg at the time when Alsace was part of the German Empire), Wilhelm Stieda (1852–1933), devoted his thesis to gathering materials for an empirical study of the ratio of the sexes at birth (Stieda, 1875). The master very quickly commented on the pupil's work, in an article where he clarified the state of his thinking (Lexis, 1876); he reproduced this text at the end of his career without really altering it (Lexis, 1903). Lexis sought to explain the regularity of the ratio of the two sexes at birth not by presupposing, from the mathematical point of view, a general "greater facility" of boys (like Laplace) or a probability that could be stated *a priori* (as in Poisson), but by seeking to capture mathematically a variability of the phenomenon that would depend both on the diversity of empirical assertions and on the number of observations. Nowadays this approach seems fairly laborious, but it must be remembered that Lexis was working in Continental Europe, and well before 20th-century parametric statistics. His argument was directed at the basis of Poisson's calculation, was fed by the breadth of empirical results obtained since, and originated from a strictly mathematical thinking where the singularity of a particular observation could not be adequately explained by a claim specific to biology or to economics, for example. The particular feature of Lexis' article is that it poses the problem of the excessive mathematical simplicity of Laplace's and Poisson's thinking (Stigler, 1986). He rediscovered a vein that Condorcet and Fourier, though each in his own way, had located before him. To illuminate this through an image, we could say that Lexis wanted to show that the object, including from the strict point of view of statistical method, is situated "below" or "within" the Laplace-Gauss distribution curve. At the end of his article, he finally appealed for a physiological exploration and mentioned the "*Sadler-Hofacker'schen Hypothese*", constructing a formula for the arithmetical link between the proportion of boys to girls at conception and its counterpart at birth (§33 and §34):

$$\frac{K}{M} = \frac{A}{B} \cdot \frac{(1-\alpha)}{(1-\beta)}$$

where K and M are the numbers of boys (*Knaben*) and girls (*Mädchen*) at birth; A and B, the numbers of germs (*Keime*) of the two sexes; α and β the fractions (*Bruchteil*) that disappear during the course of embryonic ontogenesis. He then suggests that discussion of the ratio $\gamma = (1-\alpha)/(1-\beta)$ should govern the analysis of causes of the proportion of the sexes at birth.[6]

During the 1880s, and in an entirely different university context, that of physiology at Jena, Carl Düsing himself took up the question – by

[6] We should observe that Lexis's γ is in fact a measure of the probability of survival of boys compared to that of girls. It would be accessible to empirical measure when the primary sex ratio was balanced, since it would then be exactly equal to the ratio of the two sexes at birth.

commenting on the works of Girou, "Hofacker and Sadler", Burdach and Thury, for example; by positioning himself explicitly in the Darwinian landscape of comparison between species; and by bringing together tables from the extensive literature of statistical collections then available. His thesis summarized this investigation (Düsing, 1883). A first work, disseminated both in the form of an article and of a separately printed pamphlet, gave his theory of the regulation of the ratio of the two sexes at birth (Düsing, 1884). A second work, some years later, applied it to human births in Prussia (Düsing, 1890). In the first two of these works, he proposed an algebraic formula to account for his theory of regulation. The mechanism stylized in this way has been taken by various contemporary commentators as being a prototype mathematical model in biology. This is how we should read the recent translation into English of relevant extracts from "*Die Regulierung das Geschlechtsverhältnisses… [Regulation of the Sex Ratio…]*" (Düsing, 1884), which Anthony W. F. Edwards (2000) presents as a significant moment between the Darwin of the first edition of *The Descent of Man* (1871) and Fisher's *The Genetical Theory of Natural Selection* (1930).

It is true that Düsing illustrates his argument with algebraic notations, a little like Lexis. Nevertheless, everything suggests that German statisticians and physiologists in this field were unaware of one another. Düsing did not start from Lexis' article; and Lexis, republishing his 1876 article in 1903, updated it without even mentioning Düsing's publications (Lexis, 1903). There is a fundamental reason for this: Lexis wanted to examine proportions up to birth (the primary and secondary sex ratios), while Düsing endeavoured to explain the ratio of "numbers of males and females at the time of reproduction" (the tertiary sex ratio). The Jena physiologist started from reading Darwin, sharing with him a certain confusion between stocks and flows, and seeking to circumscribe by formulas the interconnected things discussed by his predecessor. But while Darwin intended to show how the concepts of natural selection and sexual selection worked, Düsing had another priority: to demonstrate a form of regulation from one generation to another. He therefore envisaged a population with an imbalance between the sexes, and its offspring after two generations. Aware of the limits of his algebraic notation, he made it clear that: "It is true that in each individual case this is subject to considerable variation, but if one wants to illustrate and calculate the total effect in an example one must naturally use the average number" (Düsing, 1884; translation by G. S. Neumann from Edwards, 2000). So here we are faced with a schema of thinking supported by an algebraic shorthand that is constructed neither from the mathematical point of view in the manner of Laplace or of Poisson, nor from the statistical point of view in the way outlined by Lexis. There is a wide gulf between laying out a formula to clarify a narrative argument – indisputably a strength of Düsing's thesis – and thinking like a mathematician, in whatever specialism.[7] His reasoning can be followed below (see Box, p. 110). It has the obvious merit of giving a formal explanation compatible with the combinatory argument touched on by Darwin in 1871 (what he called the "doctrine of chances", 1871, p. 316) and later left in shreds. Yet there is nothing to prove that Darwin really followed this line of thinking. Moreover, Düsing's schema rests on a flimsy construction: an "average" reasoning, which is exactly the same from generation to generation (although, basically, why not?), and, in particular, a pairing of mothers according to the inverse frequencies of the sexes in their progeny. Where could such a symmetry in construction actually come from? From an implicit schema of offsetting "mothers with daughters" against "mothers with sons" – that is, from a schema that balances the pluses and the minuses, in the mould of Quetelet's analogy between the variability of things and compensation for measurement errors. Here we are very far from Darwin. In other words, Düsing's schema, compatible with a regularity assessed at the level of large numbers, is constructed in such a way as to safeguard this regularity. He proposes a mechanism that ensures a balance between the possible variations, without the aid of the calculus of probabilities. It can be seen that this reasoning involved something of a *petitio principii*.

As Edwards (2000) underlines, Düsing's schema aroused the attention of some commentators, and notably that of the logician John Venn (1834–1923), who briefly described it in *The Logic of Chance* (1888). For his part, the British economist and statistician Francis Ysidro Edgeworth (1845–1926) viewed it as a pertinent example of the "elimination of chance", which he himself placed among the results of the questions raised by Lexis, offering a key to interpretation which, we have seen, it did not have in Germany at the time (Edgeworth, 1892; on this author, see Stigler, 1986, 1999). And, a quarter of a century after its first publication, the physiologist's argument appeared to be a notable but finally futile endeavour (Gini, 1908; Halbwachs, 1912).

[7] One can hope to bridge such a gulf only by reconstructing the long history of the relationships between argumentation and mathematical thinking, which would lead towards the study of the early modern formation of algebra. This would therefore take us via the conception of mathematical writing in Leibniz and in… Condorcet. On this subject, see Cifoletti (2006).

Carl Düsing's schema of regulation (1883–1884)

(1) Düsing considers a population at the time of reproduction, which is made up of x females and $n.x$ males (n is therefore the tertiary sex ratio calculated according to the ratio of numbers of boys to that of girls), where z represents the offspring. The average offspring from each female is therefore z/x, and from each male z/nx.

(2) Secondly, he considers a mother who produces *ba* boys and *a* girls (that is, characterised by a secondary sex ratio b calculated in the same way). In the second generation, by virtue of (1), this grandmother would have $ba.z/nx$ grandchildren born of her sons and $a.z/x$ grandchildren born of her daughters, so that the number of descendants is:

$$\frac{az}{x} \cdot \left(\frac{b}{n} + 1\right)$$

(3) In parallel, he considers a mother who produces *a* boys and *ba* girls (i.e., a secondary sex ratio of $1/b$). In the second generation, as before, this second grandmother would have $a.z/nx$ grandchildren born of her sons and $ba.z/x$ grandchildren born of her daughters, so that the number of descendants is:

$$\frac{az}{x} \cdot \left(\frac{1}{n} + b\right)$$

(4) He gives the ratio of the descendants of the second grandmother to those of the first (this is the formula below) and he discusses various scenarios numerically.

$$\frac{(1 + bn)}{(b + n)}$$

(5) If the tertiary sex ratio of the population is balanced ($n = 1$), the formula allows the conclusion that the descendants of the two grandmothers are equal in number, and that this equality is independent of the imbalanced secondary sex ratio that is characteristic of them (this can be checked by an elementary calculation, such that if n is close to 1, the ratio of the two yields will be too – in fact, it will be even closer).

(6) If the initial population was very imbalanced (he takes the example $n = 2$, with twice as many males as females at the time of reproduction), and if, for example, the first "grandmother" produced 3 boys for 1 girl, then the number of her grandchildren would be 5/7ths of that of another grandmother who produced 1 boy for 3 girls. This suggests a sort of regulation.

In his thesis, examined at the University of Bologna in 1907 and published a year later, the young Italian statistician Corrado Gini took the time to discuss Düsing's works in a chapter entitled "The regulating mechanisms of sex ratio at birth" (Gini, 1908, pp. 333–339) – in order to dismiss them. Gini went on to become Italy's most important statistician. His career began at the time when statistical mathematics was managing to free itself from its old dependencies on political economy and administrative statistics (Prévost, 2002). He reached maturity under Fascism, embodying specialized expertise and its uses under this regime. His activities and his influence, in biometry and mathematical statistics, were to last through the Second World War and into the 1960s (Ipsen, 1996; Israel & Nastasi, 1998). We owe to him some of the particularly subtle mathematical thinking produced by 20th-century statistics (Barbut, 1984). The 1908 work was ambitious. Gini's intention was to propose a theory of the formation of the sexes, notably in order to respond to ideas that Italian Darwinism might have passed on about natural selection (ideas which were false, incidentally, cf. Landucci, 1996), to physiological concepts of regulation like Düsing's, and even to theories of heredity put forward by various authors in relation to the formation of the sexes. Particularly well-informed about the Continental European literature of statistics and biology, though paying less attention to English-language works which could have alerted him to fresh post-Galtonian developments, in his thesis Gini demonstrated a formidable capacity to put earlier works to the proof and to propose new calculations. Thus, for example, Chapter X, on individual variability, pioneered methods for systematic comparison of observed frequencies and theoretical frequencies and even a use of the median (1908, pp. 371–393). In 1908, Gini – like Lexis, who had gone before him along this path, but carrying a different load – handled the variability of things like a mathematician, managing to free himself from the Quetelesian mode of thinking. It was only later that he was to encounter the rise of English biometry.

The starting-point of his theory of the production of the sexes was his assertion of the relevance of a strictly statistical approach. Through the use of the most varied sources – somewhat like Darwin, but inverting the argument on the systematic nature of sources in the human case – he intended to establish that the stability of the sex ratio at birth was the distinctive feature of the human species. In his eyes, other species, insofar as they had been measured, showed greater internal variability of the same index. He then explained this regularity in human activities. Gini reduced this stability, quasi-perfect from the point of view of the calculus of probabilities (p. 136), to environmental conditions peculiar to the species.

He considered he had shown that variability in the environment, through the sensitivity of mothers during reproduction, was the cause of variability in the proportion of the sexes at birth. The human species, because it experienced more stable and more homogeneous conditions than others, would present greater stability in the secondary sex ratio. In this regard, Gini's thesis occupies a singular place in the history of discussion of this index: it has the particular feature of skirting round the issue of a physiological or strictly biological basis. Gini, who had a good command of Laplacian thinking and developed methods of putting empirical records to the proof, went back to Laplace and sought to integrate into his analysis the thing he had drawn from a century of research: the necessary variability of the phenomenon. He did not seek to construct a general stability by means of a physico-mathematics, a physiology or a theory of natural selection, but to construct mathematical indices suitable for measuring the degree of empirical variability and, consequently, describing the stability of the human phenomenon under consideration. In the sociologist Maurice Halbwachs (1877–1945) – a pupil of the sociologist Emile Durkheim (1858–1917) – Gini was to find an attentive reader who would take up the issue from a completely different point of view, as we shall see in the following chapter.

Compared to Düsing's and Gini's endeavours in Continental Europe, *The Genetical Theory of Natural Selection* by Ronald A. Fisher (1890–1962), which was initially published in 1930, is generally regarded as a neo-Darwinian work that goes further along the path opened by Galton, since in the meantime British biometry and eugenics had developed considerably (Pearson, 1978; Porter, 2004; Stigler, 1986) and Georg Mendel (1822–1884) had identified his combinatory theory of heredity (Conry, 1981; Norton, 1981). As we have already said, it was also with Fisher that a statistical methodology took shape. This was to govern the future of statistics in biology, economics and social sciences (Fisher, 1925). Several authors have analysed the logic of the principle that is nowadays associated with his name (Edwards, 1998; Hamilton, 1967; Hardy, 2002; Majerus, 2003; Williams, 1975). It is summed up in this conclusion: "It is shown that the action of Natural Selection will tend to equalize the parental expenditure devoted to the production of the two sexes" (Fisher, 1930 [1958, p. 162]). This was extended several decades later by the biologist Robert L. Trivers and the mathematician Dan E. Willard: "In species with a long period of parental investment after birth of young, one might expect biases in parental behaviour toward offspring of different sex, according to parental condition; parents in better condition would be expected to show a bias toward male offspring." (Trivers & Willard, 1973, p. 91). Although both of these publications nowadays serve as landmarks on the horizon of the collective memory of biologists, it must be said, now that we have covered three centuries, that a trend towards adjustment is not the exclusive preserve of a concept based on natural selection and that the cause of the excess of boys at birth in the human species has gradually left the sphere of physiology and extended to the spectrum of economic and social conditions.

Before setting out along the path of research into human sex ratio at birth in the history of sociology – which will be the object of our next chapter – we should observe that, by these very different routes, our study of this cause has succeeded in highlighting phenomena of underlying adjustments towards an equilibrium, and even of oscillatory logics (from Condorcet to Fisher), as well as forms of regulation of all kinds (dynamic regulation of moral phenomena in Condorcet; a trend specific to natural selection in Darwin and Fisher; a schema of numerical regulation in Düsing; and reduction in the variability of the environment in Gini).

Research subsequent to Darwin's books turned the issues of the early 19th century – when a strictly mathematical concept (Laplace and Poisson) had been opposed to an empirical investigation directed at particular causes (Fourier, Quetelet, Babbage) – completely upside down. As soon as the variability of the phenomenon had entered a conceptual framework emancipated from Quetelet's conception, where it had been the equivalent of an error dispersion, statisticians – who were mathematicians in the manner of Galton, Lexis or Gini, or else physiologists like Düsing – sought the elimination of chance and the explanation of variability by rescuing the regularity of the sex ratio from Laplace's statics, giving it the dynamic elements summed up by Fisher's principle, but also opening a new gulf between mathematicians in the style of Poisson and statisticians like Lexis or Edgeworth. Two early-20th-century French authors made these two positions precisely clear. Thus, in 1909, the mathematician Émile Borel (1871–1956) wrote:

> "There is thc same difference between these statistical probabilities and probabilities that are abstractly and rigorously defined as between the figures studied in geometry and the fairly rough representations of them that are encountered in nature: between a sphere and an orange." (Borel, 1909).

The dissatisfaction that might be shared by statisticians, biologists and sociologists on reading this passage from Borel can be measured by looking carefully at the attention that the young sociologist Maurice Halbwachs accorded to Düsing, whose narrowness and schematism he deplored in his thesis on Quetelet's "average man", a work in which he also demonstrated a solid knowledge of the works of Stieda, Lexis, the French biologist Yves Delage, Gini and Borel:

> "Therefore this is a constant cause [i.e. self-regulation], about which there is nothing accidental or indeterminate, and which explains the re-establishing of balance through the very fact of imbalance. As it [i.e. the constant cause] comes into play as soon as the balance is disturbed, the theory based on the calculus of probabilities is valid only in the hypothesis where balance is maintained: it [i.e. this probabilistic theory] implies that there is already equal distribution of the sexes[.] Yet it is just this that it [i.e. this probabilistic theory] claims to explain: in cases where it is of no use, it becomes wrong." (Halbwachs, 1912, pp. 90–91).

4. INTERLUDE: IN WHICH WE PUT PAID TO SOME COMMONPLACES

Here we come to a fork in the road: before setting off to follow one of the other paths of human sex ratio at birth, the pathway of the social sciences, it seems useful to take our bearings in relation to some of the commonplaces whose traces we have spotted from time to time – since six months never go by without some scientific journal publishing an article on human sex ratio at birth which, with all the appearance of novelty, just goes back to variants of the analytical schemas of old.

In the literature of biology published nowadays, evaluation of sex ratio at birth relates to two different objects. One is the primary sex ratio or, more rigorously, the study of factors that may contribute to determining or differentiating sex. The other is the secondary sex ratio, which is measured through real births where multiple factors combine.

In the first case, the variability of primary sex ratio according to species is the object of much research (Hardy, 2002; Majerus, 2003). As far as the human species is concerned, even a proportion of 170 boys per 100 girls ($100.M/F$) has been envisaged, which is a ratio of 63% (M/N) (Pergament, Todydemir, & Fiddler, 2002). This estimate is highly speculative. In fact, humanity does not seem to be characterized by great variability in environmental factors; moreover, it has been observed that these may play diverse roles in certain species, as in the case of temperature conditions for determining and differentiating sex (Strüssmann & Patiño, 1999). As for strictly genetic hypotheses, to date these still remain fairly unconvincing (Styrna, 1999). In addition, although there has been experimentation in the area of artificial reproduction with screening processes based on the difference in velocity between spermatozoids which carry the X or Y chromosomes, and this seems to have resulted in variations relevant to the probability of one or other of the two sexes, there is nothing to guarantee that this differential in velocity plays a part in the natural process of fertilization (Ericsson & Ericsson, 1999). Thus, balance between the two sexes at fertilization remains the most solid hypothesis (Cavalli-Sforza & Bodmer, 1999, p. 654).

In the case of the secondary sex ratio, it may be imagined from a very general point of view that the human species itself must regulate its own reproduction through the proportion of the sexes at birth, while anticipating the constraints that would be brought to bear on subsequent generations (Lummaa, Merilä, & Kause, 1998). In this argument, there is something of a teleology of the preservation of the species, which gives it a clearly metaphysical turn. Darwin, as we have seen, did not ask so much from it. Nor was he unaware that the conditions for reproduction of the next generation depend only partly on the secondary sex ratio; moreover, he accepted that species do disappear... By contrast, as we can see, the hypothesis of anticipation at the scale of the species seems to operate like Süssmilch's providence. There is nothing to guarantee that the distribution of the sexes at birth should obey such a powerful optimum, even a modernized one.

Let us now turn to four variation criteria found in the 19th-century body of work, which can also be found from time to time throughout the following century: the excess of boys among stillborn babies, which may be said to provide evidence of the structural fragility of this sex; the hypothesis of a difference attributable to different races; the greater proportion of girls among illegitimate births; and finally, the differences between urban births and those taking place in the countryside.

The heads of offices of statistics in late-19th-century Europe, as we have seen, had some knowledge of the weaknesses of registration in their day, and of the limits of their collective action to improve it (see previous chapter). This did not prevent some of them – precisely because they were the best informed – from publishing reference articles based on studies of these compilations: for example, in France, the statistician Louis-Adolphe Bertillon published a long review article, "Stillbirths", in Dechambre's

Dictionnaire encyclopédique des sciences médicales (Bertillon, 1876). For more than a century, this text provided a model for demographic studies of the comparative rate of stillbirths. It reinforced what Babbage had noted fifty years earlier: that boys were said to appear significantly more often among stillbirths than among live births. During some recent research, we were able to review the complete register of births for France for the years 1975–2004, which included almost 22.9 million registered births (Brian & Jaisson, 2007). Collection procedures and nomenclatures for "non-live" births in France were revised only recently, in 1998 (Beaumel, Désesquelles, Richet-Mastain, & Vatan, 2004). It follows that a persistent commonplace – due to the weakness of sources – must also be revised: nowadays, a procedure that is as rigorous as it can be gives sex ratios for live births and "non-live" births that are impossible to distinguish (see Table 3, below). This means that the supposed "excess of boys" among "stillbirths" must be attributed to the vagueness of earlier forms of declaration for registration purposes. Bearing in mind the assertions of several historical demographers (Armengaud, 1973; Séguy, 1997; Vallin, 2002) – which are reinforced by the study of cultural biases which, in countries where the statistical apparatus

Table 3: Non-live births in France (1975–2004)

Before procedures were revised	"Live" births	Births declared as "non-live"
From 1975 to 1997 (N)	17 327 029	145 369
Ratio of "non-live"/"live"	–	0.84%
Boys (M)	8 884 238	76 876
Sex ratio (M/N)	51.27%	52.88%
95% confidence interval	0.02%	0.26%
Maximum 95%	51.30%	53.15%
Minimum 95%	51.25%	52.62%
After procedures were revised	**"Live" births**	**Births declared as "non-live"**
From 1998 to 2004 (N)	5 536 410	37 056
Ratio of "non-live"/"live"	–	0.67 %
Boys (M)	2 835 163	18 844
Sex ratio (M/N)	51.21 %	50.85 %
95% confidence interval	0.042 %	0.519 %
Maximum 95%	51.25 %	51.37 %
Minimum 95%	51.17 %	50.33 %

Source: I.N.S.E.E., individual declarations of births, 1975–2004 (data processing: Brian & Jaisson, 2007).

is less systematic than in France, favour the declaration of boys among stillborn infants (W.H.O., 1978, 2006) – it is reasonable to think that, at the times when the idea of a greater differential rate of boys among stillbirths was formed, the probability of a stillborn baby being omitted from baptismal registers or even civil registers was greater for a girl than for a boy. Nowadays, the intensification of neonatal medicine and the systematization of declaration procedures combine to reduce the marginal space into which this disparity in treatment so far insinuated itself that, in the past, numerically relevant deviations appeared. It follows that, in these matters, a difference in morphological constitution between the sexes, or even a greater structural fragility of boys, need no longer be invoked as a cause.

In France, it is not current practice to record ethnic or racial identity. This is the object of vigorous debate and experimental exploration (Simon & Clément, 2006). But in the United States, the apparatus of official statistics, at the level of each State as well as the Federal level, produces a number of reference tables drawn up according to criteria of race (for the history of this particular feature, see Schor, 2001; Zuberi, 2001). So this makes it possible to distinguish between births, and sex of births, according to the mother's recorded "race". Statisticians who comment on these figures regard these registration categories as bearers of an intrinsic meaning. This sentence is an example of that: "The United States sex ratio at birth had three significant transitions from 1940 to 2002 (1942, 1959, and 1971). White women were the only race group to have any significant changes in the sex ratio between 1970 and 2002 (1972, 1976, and 1988)." (Mathews and Hamilton, 2005).[8] As soon as one moves away from this implicit naturalization of

[8] The authors of a second report (Martin et al., 2005) do not hide their difficulties, and they acknowledge some linguistic conveniences: "Race and Hispanic origin are reported independently on the birth certificate. In tabulations of birth data by race and Hispanic origin, data for Hispanic persons are not further classified by race because the majority of women of Hispanic origin are reported as white. Most tables in this report show data for these categories: white total; non-Hispanic white; black total; non-Hispanic black; and Hispanic. Text discussions are for non-Hispanic white, non-Hispanic black, and Hispanic mothers wherever measures for these groups are available. Data for American Indian and Asian or Pacific Islander (API) births are not shown separately by Hispanic origin because the majority of these populations are non-Hispanic. Data are also presented for four specific Hispanic subgroups: Mexican, Puerto Rican, Cuban, and Central and South American, and for an additional subgroup referred to as 'other and unknown Hispanic'. Text references to black births and black mothers or white births and white mothers are used interchangeably for ease in writing" (p. 3). On the history of racial nomenclatures in the United States, see Schor (2001) and Zuberi (2001).

nomenclatures, the figures they offer (see Table 4, below) in fact tell us that, in the United States today, births to mothers classified as "whites" present a sex ratio comparable to births in most European countries, or indeed in Brazil, Morocco or even Japan (see Table 5, in Chapter 6, p. 151–152). As for the children of mothers classified as "blacks", they are distributed at birth according to proportions comparable to those of births in Guatemala or Ecuador, while Mexico is characterized by an even lower level of secondary sex ratio, the lowest to be found in the figures compiled by the UN. Taking into account that, from a historical point of view in countries that are today the richest and from a comparative point of view in countries across the present-day world, the general level of the proportion of the sexes at birth depends on a combination of three factors (mean fertility level, extent of minimum health and social security coverage, conditions of access to sophisticated neonatal medicine), the difference observed through race nomenclatures in the United States seems to express divergences in social conditions and not a strictly biological phenomenon (Brian & Jaisson, 2007). What is

Table 4: Sex ratios at birth in the United States according to record of mother's race (1993–2003)

Sex ratio (M/N)	United states	Mother's declared race	
	(100.0%)	"White" (78.9%)	"Black" (14.7%)
1993	51.22%	51.31%	50.69%
1994	51.17%	51.24%	50.69%
1995	51.20%	51.27%	50.76%
1996	51.15%	51.22%	50.69%
1997	51.17%	51.22%	50.76%
1998	51.15%	51.20%	50.84%
1999	51.20%	51.27%	50.76%
2000	51.17%	51.22%	50.76%
2001	51.12%	51.15%	50.79%
2002	51.17%	51.22%	50.79%
2003	51.20%	51.22%	50.88%
95% confidence interval	±0.05%	±0.06%	±0.13%
Average annual births (N)	4 089 950	3 225 848	599 847

Sources: The figures for 1993 to 2002 come from Mathews and Hamilton (2005); those for 2003 from Martin, Hamilton, Sutton, Ventura, Menacker and Munson (2005). The sex ratios (M/N) shown in this table are deduced from those published in the sources (100.M/F).

more, in 2003 (Mathews & Hamilton, 2005), among mothers classified as "blacks", 68.2% of births were to unmarried couples. The equivalent figure for mothers described as "whites" was 29.4%. Thus, the differences between the two racial categories may also fall under the heading of a different issue: births outside marriage are more often female than other births. Therefore the argument of variability of secondary sex ratio according to race does not stand up.

But what do we know nowadays about this distinction between "legitimate" births and those often described as "natural" or "illegitimate"? Instead of the language of a 19th- or early-20th-century author, we could ask what the distinction is between, on the one hand, the offspring of official couples and, on the other hand, the children of unions not recognized in the eyes of church or state. Like other contemporary authors, Darwin ascribed this conventional criterion to probable differences in the specific conditions of the pregnancy and delivery, which might have had particular effects according to the sex of the newborn infant (Extract 5, Appendix B). Be that as it may, this was then a numerical fact that applied very generally to many countries: illegitimate births were relatively more often of girls than of boys. Next there came various "physical" interpretations of this difference. But in comparing the French case a century later, one has to consider the conventional nature of the criterion more seriously. From 1841 to 1905, among 60,199,700 registered live births, there were 4,664,200 illegitimate births, which is 7.7%. In the first case, the proportion of boys was 51.19% (±0.02% at a 95% confidence level and taking into account the fact that some sources sometimes impose approximations to the nearest hundred). In the second, it was 50.82% (±0.10%, same conditions) (see Halbwachs, 2005, p. 197, according to figures compiled from the French birth registers published by the General Office of Statistics for France). From 1999 to 2004, there were 4,776,588 live births, and among them, 2,148,656 births outside marriage, being 45%. The proportion of boys was then 51.21% for births to married couples (±0.05% at a 95% confidence level) and 51.18% (±0.07%, *idem*) for others (according to our studies of the I.N.S.E.E. files). Thus, now that almost one child in two is born outside marriage, not only has the criterion of "illegitimacy" totally lost the meaning that it might have had a century ago, but there is also no longer any distinction between the sex ratios of births inside and outside marriage.

But records in French birth registers of some 22.9 million live births from 1975 to 2004 allow us to go further: among children born outside marriage (6,358,240 cases), besides those that are unrecognized (1,373,320, which represents 6% of all recorded cases), we can distinguish between declarations made by both parents at the same time (2,754,959, or 12%), cases where the formalities are first carried out by the mother (625,856, or 2.7%) and those first declared by the father (1,604,105, or 7%). Although the proportion of boys among births outside marriage declared by both parents at the same time (51.27% ±0.06% at a 95% confidence level) is no different from that of births inside marriage (16,505,199 cases, of which 51.28% are boys ±0.03%, *idem*), the same does not apply to the other cases: the proportion of boys is significantly higher among newborns first recognized by the father (51.51% ±0.08%, *idem*), and much lower in those first recognized by the mother (50.80% ±0.13%, *idem*). For unrecognized newborns, the ratio is close to the latter (50.89% ±0.09%, *idem*). These up-to-date facts lead to a revised interpretation of the difference that used to appear to exist between legitimate and illegitimate births: it was simply an indicator of the greater propensity of fathers to recognize a "natural" child once the newborn was known to be male (Brian & Jaisson, 2007).

As for the variability between cities and countryside, one can guess that economic changes in European societies in the 19th and 20th centuries might have had an impact on its measure, its logic and its interpretation. Indeed, it was by this route that the social sciences, in the early 20th century, came to re-evaluate the study of the proportion of boys at birth. This is attested, for example, in the work of the Harvard economist and social theoretician of race, William Z. Ripley (1867–1941). He imagined that couples in the city were content with firstborn children even when they were girls, whereas those in the countryside would continue to reproduce until they had a boy; and he revealed that he most likely had not read Laplace when he added: "such persistence would evidently produce a greater excess of males as a natural result" (Ripley, 1908). As we shall see, the sociologist Maurice Halbwachs was to return to this topic in the 1930s, and was obviously better equipped from the point of view of the calculus of probabilities.

Chapter 5

A SOCIOLOGICAL ISSUE AND ITS PERVERSION (1898–1942)

1. COMTE: FORCLOSURE ON AN OBJECT

> "I am going to act as witness to the registration of the birth of Edouard Lefebvre's second son. There is a dreadful coldness in these acts of modern civil life. The modern family seems to me to belong to the registry and to statistics".

On 24 December 1863 – six years before statisticians met in discussions at The Hague – Jules de Goncourt recorded in his *Journal* the grip of a scientific and administrative revolution that had, as we have seen, been under way since the late 18th century. However, the international development of statistics, the resulting avalanche of numbers and the consequent rapid expansion of this kind of thinking are not enough to account for the late-19th-century formation of a particular philosophical method – the method governed by Emile Durkheim's (1858–1917) sociological doctrine – nor for its singular relationship to chance. For this, it is necessary to go back to the years of the Restoration and the July Monarchy, and to consider another protégé of the mathematician Joseph Fourier, Auguste Comte (1798–1857). Doing so will allow us to trace the beginnings of this other trajectory, which remained radically isolated from the calculus of probabilities for almost a century, until a pupil of Durkheim, Maurice Halbwachs (1877–1945), revived both a probabilistic vein of the philosophy of sciences and the issue of the regularity of excess male births.

The word "sociology" was not invented by Auguste Comte, but by Emmanuel Joseph Sieyès (1748–1836).[1] A political thinker of the same

[1] Emmanuel Joseph Sieyès – "Abbot Sieyès" – published the famous *Qu'est-ce que le Tiers État?* in 1789; he supported Bonaparte's *coup d'état* on 18 Brumaire; and he was one of the editors of the Year VIII Constitution that followed it.

E. Brian and M. Jaisson, The Descent of Human Sex Ratio at Birth, 121–143.

generation as Condorcet, Sieyès – although he was not a mathematician – nonetheless shared with the geometer a concern to promote the social art. His manuscripts on this subject use the written form *sociologie* (Guilhaumou, 2002).

It is not now known whether the word then entered circulation, or whether Auguste Comte himself reinvented it four decades later. Whatever the case, it was starting from an attentive reading of Condorcet's *Esquisse d'un tableau historique des progrès de l'esprit humain* that Comte set out to found a general doctrine of the formation of human knowledge, which governed a lot of later conceptions of the historicity of the sciences and concepts of the sciences themselves. But, although he viewed Fourier as one of his guardians, Comte relegated the calculus of probabilities to the ranks of "puerile activities" and a "frivolous semblance of mathematics" (Coumet, 2003). The conditions in which Comte's thought was shaped have already long been known: his first target was the scientific orthodoxy of the 1820s and 1830s, and most particularly the philosophy of the sciences – or what passed for it – of Laplace and those close to him (Gouhier, 1931, 1933–1941; Heilbron, 1990; Petit, 2003). It was as if the collective memory of the skill of late-18th-century mathematicians was carried forward by distinct groups of scholars and philosophers of the sciences, in tension with one another: Laplace and his followers monopolized mathematical orthodoxy in Paris; Quetelet and statisticians after him supported a scientific and administrative vulgate, vital for the construction of their empirical material; Comte and the positivists promoted a general philosophical doctrine that marked the training of later generations of scholars. These three intellectual traditions, although they were in conflict with one another, shared the same failing of attention with regard to the late-18th-century metaphysics of the calculus and its dialogue with Enlightenment scepticism. A fragmented, partial scientific memory, blind – "with one eye closed" might be preferable – to the metaphysical questions that lay at its foundations: all components of "positivism" in the broad sense of the word.

In Auguste Comte's eyes, a systematic order of the sciences – "Astronomy, Physics, Chemistry, Physiology, and finally Social Physics" (1830, p. 96; translation by Martineau, 1855/1974, p. 46) – governed the hierarchy of their difficulties and of their mutual dependence. It was from this subordination that the particular difficulty of the positive theory of political science arises. He first of all called it "social physics" (1830) and then – after the appearance of Quetelet's book (1835) under this title, which

Comte viewed as a corruption – "sociology" (1839). Sociology, in Comte's sense, was subordinated to organic phenomena. Organic phenomena were, in their turn, based on inorganic chemistry, itself rooted in physics, itself founded on mathematics).

A similar general law, known as the "three states" (See Box, pp. 123–124), represented Comte's characterization of the formation of each of these sciences (Bourdeau, 2006). Placed against the background of the academic tensions of the 1820s, and seen through the prism of his anti-probabilistic convictions, many influences and elements from the study of the regularity of excess male births can be easily recognized in each of his three states. Comte would probably have assigned the physico-theology of the previous century to the *theological state*, since physico-theology granted Providence the privilege of determining the phenomenon recorded. He would have consigned Laplace and his "greater facility" of male births to the *metaphysical state*: this facility was seen as some kind of abstract force, "capable of producing all phenomena", although only with the aid of the "frivolous" calculus of probabilities. After all, it was from the *positive state* that he expected a new social physics.

The filter of Auguste Comte's positive philosophy

Condorcet it was who grasped the general conception of the operation fitted to raise politics to the rank of the sciences of observation [...]. Not only did [he] thus conceive the method of impressing on politics a truly positive character, but he endeavoured to demonstrate the theory in the work entitled "Sketch of an historical view of the progress of the human mind". [...] *This* capital discovery has hitherto remained wholly barren [...]*

This road [i.e. "the attempt to treat social science by Mathematical Analysis, and in especial by the Calculus of Probabilities"] was opened by Condorcet, and mainly followed by him. Other geometers pursued his path and shared his hopes, but added nothing essential to his labours, at least under the philosophical aspect. *The* application of mathematical analysis is in no degree necessary to render politics a positive science. We cannot, however, stop here, for it is easy to see that such a mode of regarding social science is purely chimerical and consequently altogether erroneous. [...]*

From the study of the *total* development of human intelligence, in all *spheres*, [...] the discovery arises of a great fundamental law [...]. The law is this: that each of our leading conceptions – each branch of our knowledge – passes successively through three different theoretical conditions [...]. Hence arise three philosophies, or general systems of conceptions on the aggregate of phenomena, each of which excludes the others. The first is the necessary point of departure of the human understanding; and the third is its fixed and definite state. The second is merely a state of transition.

In the theological state, the human mind, seeking the essential nature of beings, the first and final causes [...] of all effects – in short, Absolute knowledge – supposes all phenomena to be produced by the immediate action of supernatural beings *who are more or less numerous and whose arbitrary intervention explains all the apparent anomalies of the universe.* [...] The Theological system arrived at the highest perfection of which it is capable when it substituted the providential action of a single Being for the varied operations of the numerous divinities which had been before imagined. [...]

In the metaphysical state, which is only a modification of the first, the mind supposes, instead of supernatural beings, abstract forces [...] capable of producing all phenomena. What is called the explanation of phenomena is, in this stage, a mere reference of each to its proper entity.

In the final, the positive state, the mind has given over the vain search after Absolute notions, the origin and destination of the universe, and the causes of phenomena, and applies itself to the study of their laws – that is, their invariable relations of succession and resemblance. Reasoning and observation, duly combined, are the means of this knowledge.**

* "*Plan des travaux scientifiques nécessaires pour réorganiser la société [Plan of scientific work necessary for the reorganization of society]*", 1822 (Translation by Hutton in Comte, 1877/1966, pp. 570, 577).
** *Cours de philosophie positive [Lessons in positive philosophy]*, 1830 (Translation based on Martineau's, pp. 25–26: the italicized sections represent our revisions to this classic translation).

The success of Comte's positive philosophy probably derived from this particular philosophical move, which consisted – in the most radical way, and without any real regard to an Enlightenment theory of knowledge – of proclaiming the expiry of the debates of the previous century and the futility of the scientific orthodoxy of the time, consigned to an intermediate state between the past and a supposedly new scientific spirit (Brian, 2006).

2. DURKHEIM: THE LIMITS OF METHOD

Although Durkheim retained the part of Comte's agenda relating to the need to make a general theory of the formation of the sciences conform to the delineated area of sociology, he nevertheless revised both of these, so that the new method was no longer situated after a hypothetical advancement of the sciences but at the beginnings of a theory of knowledge (Heilbron, 1990). For him, this meant asserting, in the sphere of philosophy that dealt with the moral sciences, the validity of an empirical, experimental method formed on the precedent of Claude Bernard's physiology – Bernard being the successor to François Magendie (Durkheim, 1895). Neither Comte nor Bernard could have prepared Durkheim for the need to scrutinize the calculus of probabilities: the first, as we have seen, ruled it out, while the second mistrusted "numerical methods" (Bernard, 1865, 1878). In the late 19th century, the bureaucrato-scientific revolution encouraged by Quetelet was complete, and Durkheim, like so many statisticians and economists of his generation, could handle a lot of figures without having to worry about the debates of the early decades of the century, nor even about the simplifications that, after the internationalization of statistics, had become obvious among specialists (Armatte, 1995; Brian, 1991).

The exemplar of the new sociological method was *Le Suicide* (1897; Baudelot & Establet, 1984). The distribution of the sexes at birth was not to be found in that book – although a distant but decisive trace of it came in a note: "the true founder of moral statistics is Pastor Süssmilch". However, in that section, it was seen through the filter of Quetelet's average man and Durkheim's critique of that concept (see Box, p. 126).

Durkheim's "social fact" is not the product of the average, even though averages may help us to grasp some of its manifestations. Of course, the energy that Durkheim deployed, most particularly in *Le Suicide*, in distinguishing sociology from psychology – another *terra nova* of philosophy in his day – fostered the lasting success of a commonplace confusion between "social" and "collective", to the detriment of the individual. Nevertheless, the originator of the social fact (like his successors) set out to examine it even in its most "sporadic", most singular manifestations. For Quetelet, the "average" reasoning redeemed the first appearance of the order indicated by Süssmilch. But Durkheim, because he used neither the calculus of probabilities nor Quetelet's simplification of it, had to maintain the consistency of the social fact by re-establishing its omnipotence and abandoning its only central manifestation. But in doing this, surely Durkheim did not allow Providence, which had been kept at a distance for a time but not entirely removed from the scene, to re-emerge. He himself answered this objection, following the succession of dependencies between the sciences envisaged by Comte.

The second filter: Durkheim's foundations

When Quetelet drew to the attention of philosophers* the remarkable regularity with which certain social phenomena repeat themselves during identical periods of time, he thought he could account for it by his theory of the average man – a theory, moreover, which has remained the only systematic explanation of this remarkable *property of social facts* [...].

In short, Quetelet's theory rests on an inaccurate observation. He thought it certain that *consistency* occurs only in the most general manifestations of human activity; but it is equally found in the sporadic manifestations which occur only at rare and isolated points of the social field [...].

So it is a profound mistake to confuse the collective type of a society, as is so often done, with the average type of *the individuals who make up the society* [...]. This, which is the very mistake committed by Quetelet, makes the origin of morality an insoluble problem.

* If Quetelet is the first to try to give a scientific explanation of this regularity, he is not the first to have observed it. The true founder of moral statistics is Pastor Süssmilch, in his work, *Die Göttliche Ordnung* [...] (Durkheim's note).

Source: *Le Suicide*, 1897 [1967, pp. 337, 340 and 359]; translation based on Spaulding and Simpson, 1951, pp. 300, 302 and 317: the italicized sections represent our revisions to this classic translation.

"*If we* refuse to accept that these [social] phenomena have as a substratum the conscience of the individual, we assign them another; that formed by all the individual consciences in union and combination. There is nothing substantival or ontological about this substratum, since it is merely a whole composed of parts. But it is just as real, nevertheless, as the elements that make it up; for they are constituted in this very way. They are compounds, too. It is known today that the ego is the resultant of a multitude of conscious states outside the ego; that each of these elementary states, in turn, is the product of unconscious vital units, just as each vital unit is itself due to an association of inanimate particles. Therefore if the psychologist and the biologist correctly regard the phenomena of their study as well founded, merely through the fact of their connection with a combination of elements of the next lower order, why should it not be the same in sociology? [...] But it is enough for us to show that our sociological conceptions, without being borrowed from another order of research, are indeed not without analogies to the most positive sciences." (1897, [1967, pp. 361–363]; translation from Spaulding & Simpson, 1951, pp. 319–320).

Durkheim's work, as we have seen, was as instigator of a new positive method (Besnard, Borlandi, & Vogt, 1993). He drew many of his empirical materials from those produced in the German universities in the field of moral statistics, which was dominated by the figure of Georg von Mayr (1841–1925), and from French administrative statisticians such as Alfred Legoyt (1812–1885) and Louis-Adolphe Bertillon (1821–1883). These two movements were directly responsible for continuing Quetelet's undertaking on an international level, but in particular national conditions: in Germany, in universities and regional offices of statistics; in France, outside the sphere of the universities, in the centralized bureaux of ministries and the General Office of Statistics, as well as the Office of Statistics for the City of Paris.

With Durkheim, the peculiar feature of the scientific claim of sociology did not lie in recourse to figures, nor in justifying itself. It was that, by reviving Comte's schema and defining its own empirical territory, the resistance and consistency of social facts would be noticed, preconceptions that compromised their analysis would be challenged, and it would be limited to accounting for these facts solely through the action of other social facts: "the determining cause of a social fact should be sought among the social facts preceding it" (1895 [1981, p. 109; translation by Solovay & Mueller, 1938, p. 110]). Or, in the terms used by Marcel Mauss (1872–1950) and Paul Fauconnet (1874–1938):

> "Since social facts are specific, they can be explained only by causes of the same nature as themselves. Therefore the sociological explanation should proceed by moving from one social phenomenon to another. It can only establish a relationship between social phenomena." (1901, [1969, p. 159]).

3. HALBWACHS: A PROBABILISTIC SOCIOLOGY

Among Durkheim's successors, it was Maurice Halbwachs who devoted a part of his research to exploring the distribution of the sexes at birth (Halbwachs, 1912, 1933, 1936 [2005]). Three models guided his investigation, starting with that of Durkheim himself. But, in his youth, Halbwachs also shared with his older contemporary, François Simiand (1873–1935), a solid philosophical training under Henri Bergson (1859–1941). An admiring pupil for seven years up to 1901, Halbwachs then maintained a cordial dialogue with his teacher until the end of his life. The philosopher's "sustained will" was a model, as was his rigour in the reading and criticism of older works (for example, in Bergson, 1995). However, Halbwachs did not share Bergson's later doctrine (for example, in Bergson, 1932).

His third model, to which the two described above led him, was a much older one: Leibniz (1646–1716). As a young *agrégé* – after qualifying to teach philosophy – Halbwachs took the first step in his academic career by participating in a collective scholarly project to publish Leibniz's works. Drawing on this involvement, in 1906 he published a small book that was to long remain a reference work on the subject; it was expanded in 1928 and re-edited in 1950. The combinatory notion of the social in Durkheim, the anti-mechanistic questioning peculiar to Bergson, and the tension between these two points of view, were probably what prompted the young sociologist to look more deeply into a philosophy that had accompanied the invention of the integral calculus two centuries earlier, promoted the flourishing of physico-theology in the 18th century (Rohrbasser, 2001) and governed a number of moral notions discussed in 19th-century Germany (see Box below).

A Leibnizian vade-mecum for the 19th and early 20th centuries: monadology and optimism according to Condorcet

[Leibniz] constructed the universe from simple, indestructible entities equal by their very nature. The relations of each of these entities with all the others, which with it form part of the system of the universe, determine those qualities of it whereby it differs from every other. The human soul and the least atom of a block of stone are, each of them, one of these monads, and they differ only in the different place assigned to them in the universal order.

Out of all the possible combinations of these beings an infinite intelligence has preferred one, and could have preferred one only, the most perfect of all. If that which exists offends us by the misery and crime that we see in it, it is still true that any other combination would have had more painful results. We shall explain this system which, being adopted, or at least upheld, by Leibniz's compatriots, has retarded the progress of philosophy among them. One entire school of English philosophers enthusiastically embraced and eloquently defended the doctrine of optimism, but they were less subtle and less profound than Leibniz, for whereas he based his doctrine on the belief that an all-powerful intelligence, by the very necessity of its nature, could choose only the best of all possible worlds, the English philosophers sought to prove their doctrine by appealing to observation of the particular world in which we live.

Condorcet, *Esquisse: Ninth Epoch*, Year III-1795 [2004, p. 384; translation by Barraclough in Baker, 1976, p. 227].

From his first works, Maurice Halbwachs never ceased to explore the study of statistics and of the calculus of probabilities (Fréchet & Halbwachs, 1924; Halbwachs, 1912, 1933, 1936). His work is known nowadays for two, often unrelated, aspects: firstly, his social morphology, his study of classes and his sociology of the city (Halbwachs, 1930, 1938, 1972), and secondly, his collective psychology and his theory of collective memory (Halbwachs, 1925, 1941, 1950).[2] However, an attentive analysis of the sociologist's intellectual journey shows that in fact these are two faces of the same stochastic conception of social phenomena and of a sociological method constructed in good part on the calculation of probabilities. Compared to those of his predecessors whose work he was familiar with, for Halbwachs this was a conception that went beyond the descriptive use of statistics, as in Durkheim, and the idealization of the effects of the number of individuals on forms of social groups, of which Georg Simmel (1858–1918) provided an instance in the first issue of *L'Année sociologique* (1898). It was also a response to another internal attempt to go beyond Durkheim's thinking: François Simiand's neo-positivist "phenomenoscopy" (Halbwachs, 1923; Simiand, 1922, 1932).[3]

Halbwachs' intention was to investigate social facts – which, according to Durkheim, are to be viewed as things – through thedifference between

[2] We should clarify that, in this book, we use this theory of collective memory to consider issues of scientific traditions and forms of transmission of knowledge.

[3] On this subject, see Halbwachs et al., *Le Point de vue du nombre* (1936 [2005]) and Brian and Jaisson (2005b). The works of the authors mentioned are: Durkheim, *Le Suicide* (1897), Simmel, *Soziologie. Untersuchungen über die Formen der Vergesellschaftung* (1908), and Simiand, *Statistique et expérience* (1922) and *Le Salaire* (1932). To better capture Halbwachs perspective, the work of Célestin Bouglé should be added to the list: *Les Idées égalitaires* (1899) and the critique of Georg Simmel in the first volume of *L'Année sociologique 1896–1897* (1898, pp. 152–155).

their regularity and the chance of all things. This approach may nowadays seem even more well-founded because it is on the principle of the technique of mathematical statistics testing developed since the inter-War period (Porter, 2004; Stigler, 1986). However, for the sociologist, this is not only a choice of method, but a hypothesis that relates to the social phenomenon itself: intrinsically stochastic, it lends itself to grounding in the calculus of probabilities.

In the case of social morphology (for Durkheimians, the study of collective forms), the regular features of social facts occur within groups and are expressed through the social body itself: they are produced by the combining of interactions between individuals. Durkheim's intention was to bring this combinatory logic to the collective notice of scholars, using the power of statistics to count and describe: so this was a frequentist conception of these calculations. For Halbwachs, there was no avoiding the calculus of probabilities, whose formal processes measure the difference between chance facts and social facts.

In the case of collective memory, it is the deeply intimate system of representations shared by individuals that presents regularities within the complexity of the mass of influences that each is subject to. Thus, rather than trusting chance as defined by the calculus of probabilities, we should look to chance as the principle of the subjective process by which the scholar becomes individually aware of the difference between this indecipherable complexity and the facts of collective memory (Brian & Jaisson, 2005b).

From his higher doctoral thesis on the average man (1912), through certain aspects of the textbook on probabilities that he published with the mathematician Maurice Fréchet (1924), and especially in an article presented to the Paris Statistical Society in 1933 (which fed into his contribution to the *Encyclopédie française* in 1936), Maurice Halbwachs examined the social causes of excess male births and tried to revive the discussion encouraged by Quetelet a century earlier. Then, conjecture associated with the names of Hofacker and Sadler came to his attention. He had a good knowledge of the works of the administrative statistician Alfred Legoyt, already discussed by Durkheim on questions of suicide, and he had noted that the statistician linked Quetelet's conjecture to the differences between Paris and the rest of France, observed under the Second Empire.

> "When all marriages are considered, without distinction between those registered with the civil authorities and those not, it can be seen that the absolute disproportion of age [between spouses] is greatest in this

> *département* [the Seine *département*, including Paris]; next come the towns; the rural population is in third place. – Here we find an observation that is not without importance. The official documents indicate that it is in the country areas that boys are born most, and in the Seine *département* that they are born least; therefore are we not justified in concluding from this that it is the marriages where the ages of the spouses are least disproportionate, especially in the period when these marriages may be fertile, that give rise to the largest number of male births?" (Legoyt, 1864, Vol. 1, p. 507)

Halbwachs could not miss the city's entry onto the scene of the causes of variation in sex ratio at birth: he was easily able to reinterpret Legoyt's question in Durkheimian terms, bearing in mind his own work on Paris (1909 and 1928). In those terms, the forms of urban and rural society here came into play.

Like Legoyt, other statisticians sought to put to the proof a possible dependence between parental age gap and the male birth rate, through perusing vast numbers of cases – tens, or even hundreds, of thousands. From this work, they all essentially concluded that the parents' ages, or the gap between them, had no effect on the sex of live births. Halbwachs knew two of these studies well: the thesis by Wilhelm Stieda (1852–1933, published 1875) – a pupil of the great German statistician Wilhelm Lexis (1837–1914), whom Halbwachs met in 1903, during his stay in Göttingen – and the thesis by Corrado Gini (1884–1965, published 1908), the most important Italian biostatistician of the 20th century (Prévost, 2002).[4] Throughout the 19th century, civil registration statistics multiplied, notwithstanding the doubts

[4] On his readings of Stieda and of Gini, see Halbwachs, 1912, 1933, 1935 and 1936. Stieda was, in a way, Halbwachs' immediate predecessor on the empirical terrain of Alsace, although he had worked there at the time when the University of Strasbourg was German. In Spring 1874, Stieda explained his research to Professor Wilhelm Lexis' seminars on statistics at this same university. The Professor provided a commentary on it almost immediately, with a revival of the binomial in mind (Lexis, 1876); on this subject, see Stigler (1999, pp. 87–128). The 1876 article formed part of Lexis' collection, published in 1903, when Halbwachs was in Göttingen. The sociologist commented on this volume in his higher doctoral thesis of 1912. We should add that, in Maurice Halbwachs' own publications, there is no record of Emmanuel Carvallo's book (1912), Chapter II of which looks closely at statistical method in the area of the births of the two sexes. This book, mentioned – even implicitly criticized – in Fréchet and Halbwachs' textbook (1924), seems to have been more in the mathematician's mind than in the sociologist's. Its treatment probably also related to two other factors: firstly, Carvallo's textbook dated "from before the War"; secondly, its author, Director of Studies at the *École polytechnique*, was addressing engineers rather than university academics who were graduates of the *École normale*. Be that as it may, Halbwachs, who would not have found the answers he was looking for there, ignored it.

about their sources expressed at The Hague in 1869. Between 1870 and 1910, a series of systematic studies made use of them. In the end, these exercises finally arrived back at the point marked by Poisson in 1830, although they sometimes led to some of the most highly developed thinking on the calculus of probabilities of their day (as in Lexis, 1876; see Stigler, 1999).

As we have seen, in 1908, Gini did more than simply add an extra title to this series. Gini's explicit intention was to set the seal on a new genre, that of biological statistics. Halbwachs knew the Italian statistician's thesis very well. He appreciated its technical subtleties and did not hesitate to use certain passages from it, almost word for word, in the study of the proportion of the sexes at birth that he himself published in the *Encyclopédie française* (1936 [2005]). However, he did not follow his Italian predecessor into the analogies, consolidated by Darwin, between the human species and species of animals. The distinctive feature of human beings, in the eyes of the sociologist, was that the phenomena affecting them are mediated by systems of representations (Halbwachs, 1935).

Comparison between his thesis of 1912 and the 1933 article demonstrates a decisive element that was an additional influence on Halbwachs: the experience of the Great War, at the end of which the proportion of male births had increased very markedly, with the result that – in his eyes – the variability of this index could no longer be in doubt (see Box pp. 132–133). Hence his return to the "Hofacker-Sadler law". In order to put this conjecture, so much discussed, to the proof, the sociologist used a new source: the records of allowances paid to large families in the Bas-Rhin *département*. This source, easily to hand in Strasbourg where Halbwachs was teaching, was a remnant of the social security laws of the German Empire. Thus, Halbwachs perused no less than 63,821 of these records, single-handed. This hardly needs to be pointed out: in his day, manual processing was the only option.

Halbwachs' critique of statistics

1912. – In short, Quetelet had a very accurate sense of the existence of the social laws that are imposed on the wills of individuals, and he spoke of society and of collective institutions as "things" that in some way carry within themselves the laws of their development. But, like almost all philosophers, moralists, statisticians and writers on social matters of his time, after having glimpsed this point of view, he did not adopt it. Since he found a large number of consistencies and regularities in social facts, he had no doubt that it was possible to explain them through simple laws that could be applied mechanically. Beneath collective facts, he sought individual approaches, especially of a physiological nature, which, whether similar, or in opposition, or reinforcing one another, or cancelling one another out, allow the most frequent ones – most frequent because they are responding to the simplest combinations of causes – to show through alone. [...] He believes that, in society, as in the world of living beings, the most frequent means is the best, the ideal. The basis of this conviction, is certainly, in short, a teleological or finalistic idea of the world (pp. 174–175).

1931. – In the face of those who view statistics simply as a means of description that produces piles of tables, without making any serious effort to elaborate on them (as is too often the case in Germany), and in the face of statistical mathematicians who see statistics as an application of mathematics and direct their efforts towards formulae and technical methods (seeking new indices of covariation, correlation, etc.), the French sociological school does its best to adopt and maintain a clearly positive attitude; it views statistics as a scientific instrument that must above all be well-adapted to its object and should be judged on what it yields ([2005], pp. 273–274).

1937. – Mathematical statistics has really made great strides. [... But] it has very rarely dealt with facts themselves. Or rather, facts are involved only to the extent that they lend themselves to the application of formulae. Mathematicians too readily lose interest in positive realities. Of course, mathematical simplifications enable clearer investigation of certain relationships between abstract data: but what do we gain from that, if it is precisely the essential element that is eliminated in the process? Formulae should be perfected and made more complex; several should be used at once, adapted step by step to the facts that they should be helping us to understand better. But since mathematicians can hardly be expected to become sociologists, it is for the latter to introduce a little more mathematical precision into their methods. They alone can know what they lack in this regard (pp. 143–144).

By choosing to measure the age gap to the nearest half-year, Halbwachs hoped to use tables and charts to highlight a quasi-"cyclical" dependence between two variables: parental age differences and proportion of male births (1933 [2005, pp. 393–399] and 1936 [2005, pp. 278–279]; see next Box, pp. 134–136). Although the imperfections of the curve thus drawn belonged, in his view, to the "sphere of life whose complexity most likely does not yield to the regularity of mathematical expressions", he nonetheless asserted "that the results of a game of chance would not line up in this kind of order" and that there was "certainly a relationship between the values taken by the proportion of male births and the age difference between the parents". The male birth rate was "the only factor" whose "effect we have [...] been able to recognize and measure" (1933 [2005, p. 396]).

The objective that Halbwachs set himself was to demonstrate that the ratio of numbers of boys to numbers of girls at birth did not arise by chance, and that society had an effect on it, following a more or less periodic mediation based on parental age gap. Halbwachs produced further details of this action, showing that it exerted varying degrees of influence in the countryside or in the towns, in wartime or in peacetime, among married or unmarried couples. In doing this, he explained variations found and confirmed both by him and by his predecessors.

The variability of the sex ratio

1933. – This is the first time since male and female births have been recorded in France that such a marked variation has been discovered. The only comparable irregularity, which appeared 116 years earlier, was a great deal smaller.

If, in ten or twenty centuries, all memory has been lost and no trace remains of the historical, political and military events that took place in France in the 19th century and the first quarter of the 20th century, and yet a statistical report is found showing, year by year, the ratio of births of boys to births of girls throughout that period of 130 years, it would have to be assumed that in around 1802–1803 – in those years themselves or in the years just before them – and, similarly, in around 1918–1919, some major upheaval must have taken place in the physical or social environment, since the balance in the ratio of births of the two sexes, usually maintained from one year to the next, was so clearly disrupted. It is natural that we should immediately think of the wars of the Revolution and of the Empire and also of the Great War that ended in 1918 ([2005], p. 384).[5]

[5] The enthusiasm expressed by Halbwachs was fuelled by reading a passage from Leibniz's *Nouveaux essais sur l'entendement humain* (1704), in which he envisaged that general truths may be conceived and conveyed without the help of verbal expression: "I believe that other marks could also produce the same result – the characters of the Chinese show this. And we could introduce a *characteristica universalis* – a very popular one, better than theirs – if in place of words we used little diagrams which represented visible things pictorially and invisible things by means of the visible ones which go with them, also bringing in certain additional marks suitable for conveying inflexions and particles. This

1936. – This is probably the first time that it can be demonstrated in a rigorous manner that variations in the ratio of male births are not simply random ([2005], p. 276).

1912. – The sex seems to depend to some extent [...] on the size of the parental age gap. Yet, within society, laws and opinion determine the lower age limit and the acceptable age gap for marriage. The age ratio of men and women able to procreate is thus explained by social causes (p. 167).

1936. A route towards a solution. – Thus, the numerical ratio of the sexes at birth is not simply a curious fact. It poses very big problems: not only the problem of the biological

---

would at once enable us to communicate easily with remote peoples; but if we adopted it among ourselves (though without abandoning ordinary writing), the use of this way of writing would be of great service in enriching our imaginations and giving us thoughts which were less blind and less verbal than our present ones are. Of course not everyone knows how to draw, so that apart from books printed in this manner, which everyone would soon learn to read, some people would only be able to make use of this system by printing of a sort – by having engravings ready to use for printing the pictures on paper and then adding the marks for the inflections and particles by pen. But in time everyone would learn to draw during childhood, so as to be able to take advantage of this pictorial symbolism; it would literally *speak to the eyes*, and would be much liked by the populace. In fact peasants already have almanacs which wordlessly tell them much of what they want to know. And I remember seeing satirical broadsheets, in copperplate, which were somewhat of the nature of puzzles, containing inherently significant pictures mixed with words; our letters and Chinese characters, on the other hand, are significant only through the will of men (*ex instituto*)" (Translation based on Remnant and Bennett, 1981, pp. 399–400: we have restored Leibniz's Latin phrase at the beginning of this scholarly translation). Given that *Nouveaux essais* went through several editions between 1866 and 1900, this passage was in effect summed up by Halbwachs, talking about the *characteristica universalis*, in his *Leibniz*, 1906, pp. 20–21 (1928, 1950, pp. 54–55). Leibniz's work, actually published in 1765, found a similar echo in Condorcet's *Essai d'une langue universelle*, a long fragment that remained in manuscript form during Halbwachs' lifetime: "But this hieroglyphic writing may be used to advantage for another object that deserves our attention here [...]. Is it not possible that a global cataclysm, while not annihilating humankind completely, swallowing it in the eternal abyss with the monuments it has raised, might nevertheless destroy the arts and sciences, and their fragile repositories, and even the languages now known. Would this writing not then be a means of preventing a part of the losses to which this revolution would condemn humankind, of erecting more lasting monuments to the sciences which only a total cataclysm of the earth could take away" (*Tableau historique*, 2004 edition, pp. 957–1014, extract from p. 995). The collective memory of Leibniz's conceptualization was therefore good, and a new confirmation of it may be found here, an effective mediation between the most advanced – but also the least known – of Condorcet's reflections and Halbwachs' discoveries: thus, this "philosopher's memory" operated like a sort of time machine, which, in the 20th century, carried the sociologist back to the same environment as his predecessor 140 years earlier, without his even knowing Condorcet's relevant writings.

determination of sex, but also the demographic problem of the influence of marriage regimes, the strength of marriage trends and customs affecting those trends on the composition of groups and on the proportion of men and women, of boys and girls. It has long been believed that births of the two sexes were in a balanced ratio as a result of the laws of probability. But, if births were comparable to chance facts, then one of the essential functions of organic life would be subject to purely mathematical or mechanical laws. Of course, in a sense, it is a matter of chance whether a boy or a girl is born, [...][6] but a given age distribution within a group is not a matter of chance; and if this distribution explains, at least in part, the numerical ratio of the sexes at birth, then the birth, in a given society, of a given proportion of boys is not a matter of chance either (1936 [2005], p. 281, drafted in 1933, [2005], p. 400).

Halbwachs' biological key to the social in 1933

Ratio of male births to female births according to parental age difference

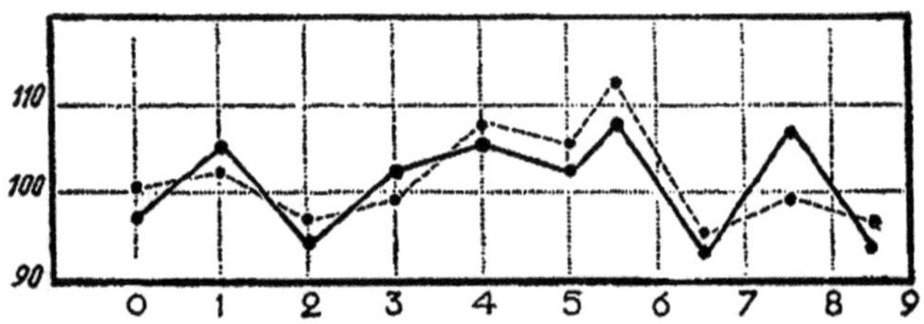

The horizontal divisions represent parental age differences, in half-years. The vertical divisions (90, 100 and 110) represent numbers of boys per hundred girls. In order to validate his solution, Halbwachs divided 42,918 births, which took place in Bas-Rhin between 1925 and 1930, into two sub-groups according to the arbitrary order in which they had been recorded (26,499 and 16,419 births respectively).

Figure published in 1933 [2005, p. 398] and again in 1936 [2005, p. 278] (Courtesy of I.N.E.D.). The title and legend we use here are those of the originals.

[6] Halbwachs includes the following excursus: "and there is perhaps a profound symbolic truth in the old notion that births were related to the conjunction of the stars". Laplace, Quetelet and Durkheim are here in the background, and this "profound symbolic truth" relates to the analogy between two combinations of chances, general laws and subjective lack of comprehension: "the totem of the child is not necessarily either that of the mother or that of the father; it is that of a mythical ancestor who came, by processes which the observers recount in different ways, and mysteriously fecundated the mother at the moment of conception. A special process makes it possible to learn which ancestor it was

Neither the Paris Statistical Society, before which Halbwachs presented his work in 1933, nor the columns of Volume VII of the *Encyclopédie française* was the appropriate place for a discussion of the anthropological bases of Leibniz's metaphysics as renovated by sociology, young as it still was. Nevertheless, here he was considering the relationship of the individual to society according to the relationship of monads to their optimal combinations. This may have been his train of thought, but he did not confuse monadological theology and society, any more than a physicist – as D'Alembert had known – or an economist would confuse them when carrying out an optimization calculation. As for the object itself (the proportion of the sexes) Halbwachs situated his thinking resolutely after Laplace's and Poisson's mathematical constructions, after the probabilistic clarification that resulted from them, after the avalanche of official statistical figures encouraged by this clarification, and finally after Gini's biostatistical investigations.

Halbwachs effected a decisive reversal of 19th- and early-20th-century ideas. The phenomenon was no longer characterized by a uniform probability that would measure the frequency of observations, where the deviations so often noticed were purely a product of uncertainties in measure and in method. This phenomenon is now stochastic in nature ("it is a matter of chance whether a boy or a girl is born"). It is estimated through its frequency, and the periodical variation in this frequency – which is certainly very minimal in amplitude – is the trace of the social fact.

> "From the time you arrived at the *Lycée Henri IV*, I saw in you the philosopher you have shown yourself to be in your books, a philosopher who would make a major contribution to sociology and give that science greater precision."[7]

With these words, written in December 1939, Henri Bergson congratulated his former pupil on becoming a professor at the Sorbonne. Indeed, the most admirable features of Halbwachs' research related to this methodological tradition in philosophy: its conceptual development, putting its metaphysics to the proof – vital to sociological theory

---

and to which totemic group he belonged. But since it was only chance which determined that this ancestor happened to be near the mother, rather than another, the totem of the child is thus found to depend finally upon fortuitous circumstances" (Durkheim, 1912 [1979, p. 150]; translation by Swain, 1915/1965).

[7] Letter from Henri Bergson to Maurice Halbwachs, dated 1 December 1939, autograph manuscript, IMEC – *Institut Mémoires de l'édition contemporaine*, Halbwachs Archive, HBW2-A1-04.1.

in the tradition of Durkheim – and its renewal, constantly fuelled by an unremitting quasi-experimental effort.

But the nodal empirical argument in his proof of a "cyclical" dependence between parental age gap and the proportion of the sexes at birth does not hold good... Using the methods then available, Halbwachs divided his batch of tens of thousands of records at random, and observed isomorphic fluctuations over the sets thus formed. Unfortunately, he did not consider the binomial confidence intervals, which are much easier to control nowadays because the use of statistical tests has become very widespread, especially since the growth in electronic calculation. Because this was the case, if we start from Halbwachs' compilations and tables, we have to recognize that the fluctuations he found remain roughly the same as the amplitude of the confidence interval around the sex ratio for births grouped according to parental age gap. Moreover, if other individual sources are mobilized to be processed using Halbwachs' approach – for instance part of a reference investigation in the field of historical demography (the Henry survey, see Séguy, 2001) – it becomes clear that no "cycles" can be found to match those he thought he had established. Thus, the "Hofacker-Sadler law" is not merely Quetelet's myth, but, in addition, empirically futile (see Halbwachs, 2005, pp. 172–175; see also Brian & Jaisson, 2007).

The most important thing in Halbwachs' work on sex ratio, therefore, is that he clearly posed the question of the nature of the phenomenon, which is both stochastic and social. Although the presence of the works of Durkheim, Comte, Quetelet, Fourier and Laplace had a filtering effect on this shifting yet recurrent picture – and despite the vicissitudes of the collective memory of Condorcet's conjecture, for all that this reformulation is preferred – Halbwachs saw the stochastic nature of social phenomena and revived Condorcet's idea that "humankind" could intervene in probability in relation to the sex ratio (expressed as a "means of producing male or female children, with a certain probability"). Condorcet, as we have seen, had sought to free himself not only from Süssmilch's theology but also from the way Laplace had conjured it away using mathematics. Similarly, Halbwachs, in carrying sociological questioning through to its conclusion, was revising the numerical immanence part of the social fact in Durkheim's conception and correcting the most widely accepted statistical conclusion: that the sex of a newborn child was a purely chance fact. Thus, at the scale of his work as a whole – as well as of a three-century survey of the regularity of birth figures – Halbwachs was replying to the 18th-century providentialist notion, as well as to 19th-century ideas, either laplacian or positivist.

4. THE WORST CASE TEST

Maurice Halbwachs' research on the proportion of the sexes at birth was published in Volume VII of the *Encyclopédie française*, which appeared in 1936.[8] This volume was envisaged by its editor, the historian Lucien Febvre (1878–1956), and by the principal authors whom he recruited, the ethnologist Paul Rivet (1876–1958) and the zoologist and anthropologist Henri Neuville (1872–1946), as a gesture of political responsibility by competent scientists faced with the rise of the apparently scholarly arguments being trumpeted by propagandists of fascist policies, most especially in Germany and Italy. As we know, the governments of these two countries mobilized entire areas of biology, anthropology and demography in order to justify their doctrine. This state of affairs, as we also know from several recent studies, profoundly distorted the conditions in which scholars active in these spheres worked (Ferdinand, 1997; Gutberger, 1994; Ipsen, 1996; Israel & Nastasi, 1998; Mackensen, 2002, 2004; Tooze, 2001; Wietog, 2001). The part of the 1936 volume where Halbwachs included his results on births aimed to provide a comparative demography of humankind to an audience that was assumed to have a close interest in things of science and culture, in order to use education to thwart the simplistic propaganda of the day and its echoes in the press or in public opinion.

During these same years, German statisticians from traditions bearing the stamp of Quetelet and von Mayr were at loggerheads in a political and scientific rivalry that provided the context for a partial circulation of the most statistical aspects of Halbwachs' research on the formation of couples and on births. However, the way it was received in Germany seems to have had no resonance in France. Scholars in Paris at that time were profoundly unaware of the forms taken by actual relations between politics and social sciences in the Germany of the Third Reich: the study of these relationships came seventy years later.

Each of the three German statisticians who read Halbwachs in the 1930s offers a somewhat edifying case to the social sciences record book. One, Roderich von Ungern-Sternberg (1885–1965), was an opponent from within, locked in a dialogue of the deaf by political configurations on both

[8] The following pages summarize the critical introduction that we edited for publication in the recent edition of Halbwachs et al., *Le Point de vue du Nombre 1936*, 2005. The elements that most particularly concern the reception of Halbwachs' research in Germany come from the article by W. Gierl and É. Brian in that volume, "*La réception de la démographie de Halbwachs sous le IIIe Reich*", pp. 131–148.

sides of the Rhine and by forms of intellectual exchange between the two countries. A second one, Philipp Schwartz (1887–1934), was heir to the Bavarian school of statistics of Georg von Mayr and his successor Friedrich Zahn (1869–1946). A member of the National Socialist Party (perhaps opportunistically), he was killed in 1934 when scores were being settled between sections of the Party in Munich. His death enabled the Berlin statistician Friedrich Burgdörfer (1890–1967), a Nazi by conviction and a zealous propagandist, to rise to the top of the Bavarian Office of Statistics. The third, Richard Korherr (1903–1989), was another spokesman for the Nazi view of statistics, a visionary of the power of figures, equally active in the specialized office at Würzburg in Bavaria and at the very heart of the Nazi apparatus: he officiated in Berlin as Inspector of Statistics under the head of the SS, Heinrich Himmler. We now know that it was in this capacity, in Spring 1943, that he edited a statistical report on the progress of the resolutions taken by the Nazi government at the Wannsee Conference in January 1942, planning to exterminate European Jews.

At the height of the War and in the context of battles inside the SS for control of the technology of the statistical calculators produced by the Hollerith company (later to become IBM) and by its competitor Powers, Korherr made use of the Halbwachs article published in 1933.[9] Between May and December 1942, he systematically studied the frequency of births of boys and girls within SS families, processing the internal file kept by this organization. Through a paper that he sent in May (published in Halbwachs, 2005, pp. 156–162) and through later correspondence, he hoped to draw Himmler's attention to the efficiency of more traditional statistical methods. In short, Korherr took a banal stand in a specific bureaucratic battle, but in extreme political and military conditions: he was manoeuvring in response to one of his boss's preoccupations – an anxiety to renew the strength of his forces, then seriously affected by the War.

In the background to this paper lay the technical idea of facilitating the birth of boys in supposedly ideologically and racially pure families. This was rooted in the institutional practices established in Germany from 1936, planning for the selection and racial protection of newborn babies in *Lebensborn-Heime* [Fount of Life Homes] maintained by the SS, as well as in other practices regulating the formation of couples in this organization.

[9] On this political and technical context, see Edwin Black's book (2001), which is the best-informed currently available. On the episode in question, see É. Brian, "*Comment contrôler le sexe des enfants? Documents sur une obsession statistique au sein de la SS*", Halbwachs, 2005, pp. 149–168.

Totalitarianism aped whatever aspects of social mechanisms it was able to perceive.

After several pages reviewing the most varied arguments, from Aristotle onwards, about the generation of the two sexes, Korherr asserted the pressing need for new research on the link between sex of children and parental age, and the opportunity that the quality of SS "material" – that is, sources and people – represented in that regard. In the grip of a hallucination that he had cultivated since the 1920s (a vision of a sociobiological war in which German scholars would have to face determined, dangerous enemies), Korherr turned to Halbwachs' work, presented as the work of a statistician – that is, of one of his counterparts, but from another side. The paper argued in favour of sociobiological experiments whose "material" would be the "living body" of the SS, and whose regulating parameters remained to be defined. The age gap between the parents? The father's age? The mother's age? These criteria were envisaged like elsewhere the nationality and race of SS member's wives. Korherr railed against the "Hofacker-Sadler law", but without considering Halbwachs' critique, figures or conclusions. He simply promised success at the end of a vast investigative undertaking. In his eyes, as in those of his reader, Himmler, it was undoubtedly important that the father's age – that is, the age of the SS male – should remain a matter of no concern, and that control should be operated by varying the mother's age. For Halbwachs, age differences were taken as absolute values, and the two parents as symmetrical.

Like Schwartz in his review of the 1933 article (Schwartz, 1934), Korherr did not understand Halbwachs' (certainly fragile) reasoning concerning "cyclical" dependence nor the (more solid) fact that parental age gap was merely the indirect morphological evidence of social phenomena (a context of crises, urban density, diversity of family group trajectories). Schwartz's strictly statistical pre-War reading of the 1933 article, like its use by Korherr as an instrument in the internal struggles of the SS in 1942, were as clumsy as they were inappropriate. Most of the figures put forward by the SS Inspector for Statistics made only little sense, once put to the test of confidence intervals (see Halbwachs, 2005, pp. 154–155). The documents that have survived from the Korherr episode reveal an obsessive, to whom only two levels of assessment mattered: the level of the very smallest things, to be calculated using the "material" that he had to hand, and the level of his most general views, where he projected ideological and geopolitical issues into his research. There was no intermediate level in this methodological

Manichaeism: no room for the application of precisely one method, one construction of the object, in a rationally controlled way.

Among the superstitions attached to the social sciences, there is a very widespread one which holds that, as soon as a social phenomenon has been defined or a figure calculated, the exercise of excessive, even totalitarian, power will follow. This was how the expression "all knowledge is power" came to undermine the history of the sciences in the final decades of the 20th century. A puzzle arises when we compare this idea to sayings of a totalitarian nature from the 1920s and 1930s. "*Il numero come forza [The number as strength]*", proclaimed Mussolini in the Fascist press. "*Die Zahl als Machtfaktor [The number as power factor]*" – Korherr went one better in the organ of the SS Directorate.[10] How did a simplistic statement like this, already forgotten, become a semi-academic commonplace? The example of readings of Halbwachs among German statisticians under the Third Reich confirms that a figure, an index or an analysis does not move so easily from the world of anti-fascist scholars to the circles of SS chiefs. This was because the transfer took place only after the traces of facts and figures had been wrenched from their initial social context (that is, a specifically scientific one), a move which distorted them.

Halbwachs' figures on the sexes at birth presented some variations that were pertinent and others that were arguable. They supported a general thesis that remains to be put to the proof, but their simulacrum in a completely different context (although it was in keeping with certain features of the statistical methodology of that period, it was marked by, among other things, an exaggerated concern to privilege paternal age and manipulate the mother's age) radically debased what it had borrowed from the sociologist. In short, this was an ideological bid to take over an academic topic.

It would be genuinely difficult to analyse this if one ascribed enormous power to numbers. Therefore it looks as if, once the silences of the post-War period had passed, the unthinking repetition of old, hateful slogans served as an outlet for collective feelings of remorse. In order to drag ourselves out of the sleep of reason that produces monsters, we need to take a case history, to measure how far the old fascist harangues and the vulgate of

[10] Benito Mussolini, "*Il problema demografico della Nazione. Il numero come forza*", *Il Giornale d'Italia*, 27 September 1928, p. 1. Richard Korherr's article, "*Die Zahl als Machtfaktor*", appeared in the organ of the SS Directorate, *Das Schwarze Korps* (archive copy: Berlin, Federal Archives, BDC-SSO Korherr, 201-A, 1366).

scientific commentary may be expressing two forms of the same fetishism of power, and may therefore carry the seeds of the same scientific nihilism.

The historical awareness of scholars has probably become less acute since the time when Auguste Comte dreamt aloud of a politics conceived as a "true positive theory". Condorcet, Laplace, Fourier or Quetelet had not asked for so much when they expressed hopes that, for the good of the political and moral sciences, "an exact register" would be kept "of the results which the various means used have produced, and which are so many experiences made on a large scale by governments".[11] This was experimental research being conducted in the laboratory of humanity, characteristic of the particular form taken by the division of labour between administration and science, first of all in France, then in the Napoleonic Empire, then – through the repercussions of early-19th-century wars – throughout Europe, and finally in many parts of the world through the effect of the international coordination of European offices of statistics and their branches in the colonial empires.

But here the experience of humankind greatly exceeds the empiricism of the turn of the 18th and 19th centuries, leading as far as the worst instances of destruction in the Second World War. It is time now to take up again the issue of the proportion of the human sexes at birth from a radically scholarly point of view, as it was by Condorcet, Poisson or even Halbwachs.

[11] Laplace, *Essai philosophique* (1921), Vol. 2, pp. 1–2; translation by Truscott and Emory from Laplace, 1951, p. 107.

Chapter 6

A STOCHASTIC RE-EVALUATION

1. IS HUMAN SEX RATIO AT BIRTH A CONSTANT?

Is there something in the proportion of the two sexes at birth that should be seen as a constant, known to the social sciences since the 18th century? This was basically Johann Peter Süssmilch's position – and although it may be anachronistic to read him in this way, such a reading is far from pointless. Demographers nowadays share a certain hypothesis: the index is a quasi-constant, except for "an unexplained upward trend in immediate post-war periods" (Pressat, 1979, p. 187). The differences sometimes observed then become indices of an arbitrary distortion that needs to be rectified:

> "The male birth ratio (number of male births in relation to number of female births) is most often very close to 105% [51.2% boys in the total number of births]. It is one of the rare demographic parameters that are almost constant. Except where there is intentional intervention affecting the sex of the child to be born (scope for which remains fairly limited up to now), variations observed around this level of 105% are both rare and slight – to the extent that, for the well-informed demographer, any observation to the contrary first and foremost arouses suspicion of a failing in the registration of births" (Caselli & Vallin, 2001, p. 57).

However, the same index has come under the full glare of the demographic spotlight for some time. The implementation in China, from 1978, of a policy known as "the one-child policy" has had the known consequence of increasing the proportion of boys among declared births to about 55%. This situation has been the object of numerous studies, and several summaries have already been put forward (Attané, 2005; Banister, 2004; Cai & Lavely, 2003; Coale & Banister, 1994). The diversity of responses from Chinese populations in cities and in the countryside to this normative demographic policy, its regional variations, and the imperfect nature of

E. Brian and M. Jaisson, The Descent of Human Sex Ratio at Birth, 145–176.

registration figures are all well-known to specialists (Zhang, 2004). But the orders of magnitude are so large that there can be no doubt of the stark increase in the proportion of boys at birth for the whole country: to be convinced of this, the reader need only look at Figure 1, below, where the accepted estimates of mean sex ratio at birth in China are represented vertically. The inadequacy of the "one-child policy" can be measured on the same graph. Whereas the crude birth rate had been falling since 1970 with practically no impact on the proportion of the sexes at birth, 1979 marked an abrupt halt to this trend. Then a new pattern became established, characterized by a much slower rate of decline in the birth rate and a startling increase in the male sex ratio at birth.

Thus it has already been the case for several years that calculating the sex ratio of live births has enabled demographers to track a phenomenon that for a long time passed unnoticed, but becomes obvious as soon as some attention is devoted to it – the phenomenon of missing girls in countries

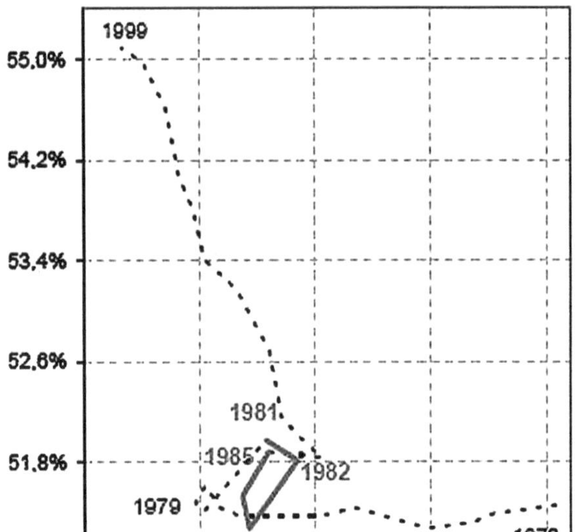

Figure 1: Birth rate and sex ratio at birth in China from 1970 to 1999. (The mean annual birth rate for the whole of the People's Republic of China is shown horizontally per thousand of the population. The proportion of male births is represented vertically. The curve follows the trajectory of these two indices from year to year.)

Source: Indices calculated from Cai & Lavely (2003) and Coale & Banister (1994).

affected by major demographic change, notably China and India. Rates of missing girls are calculated by comparison with a sex ratio at birth which is assumed to be constant. So what value should be adopted in the calculation: 51.2%? 50.0%? or even a rate appropriate to a country undergoing major demographic change? There is nothing in these examples that settles the question conclusively, even though there is no doubt that overall the sex ratio at birth does help to tell us something about a deficit in live female births (Attané & Véron, 2005; Arokiasamy, 2004; Banister, 2004; Cai & Lavely, 2003; Kim, 2004).

Let us return to one of the cases through which the idea of constancy in the sex ratio at birth gained strength among demographers – that of French civil registration. As we have seen, several of its features were analysed by the sociologist Maurice Halbwachs, who demonstrated its variability – whether or not one shares his analysis of the causes of this. Since 1975, the proportion of male births in France has been very close to 51.25%. Using registers as sources gives 168,442,939 births recorded from 1801 to 2000. Categorizing these into five-year periods gives a mean total of births in the order of 4,200,000. The limit of the 95% confidence interval of a binomial for these total numbers is then in the order of 0.05% (see Appendix D). From 1811–1815 to 1996–2000, there are several ways of enumerating six values of the estimated proportion of male births, with clearly distinct confidence intervals for these values. One series of such five-year periods is represented horizontally in Figure 2, below. For each of these periods, the probability that the sex ratio is outside the shaded interval (whose width is in the region of 4 standard deviations) is smaller than 5%[1].

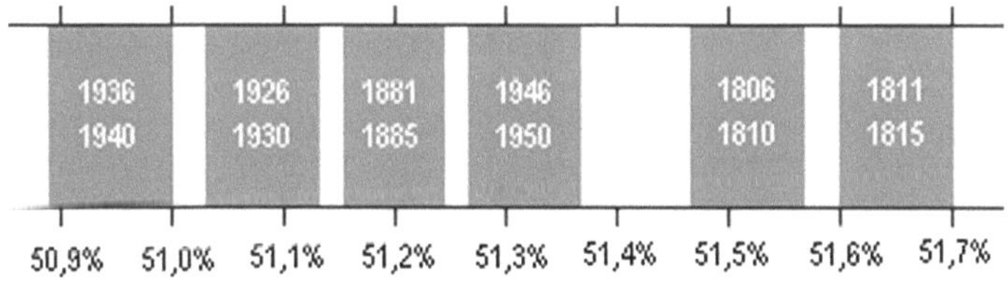

Figure 2: 95% confidence intervals for the proportion of male births among live births in France for six 5-year periods

[1] A series of six cases is shown in the table p.148 (others are possible). The sources are live birth figures from the General Office of Statistics for France and the French National

Let us assume that the proportion of male births in France from 1811 to 1950 was constant and equal to the same value *s*. Since two confidence intervals cannot overlap, two estimates of the sex ratio are at more than two standard deviations from each other. The probability of the existence of this *s* is therefore lower than 5%. For three separate intervals, the difference between the two extreme estimates is certainly higher than six standard deviations; for four, five and six separate intervals, the difference is higher than ten, 14 and 18 standard deviations respectively. Thus the probability of the existence of *s* vanishes to a value of almost nil... Therefore it is not reasonable to believe that the proportion of the sexes at birth in France in the 19th and 20th centuries was constant, and still less that it must have more or less coincided with the value 51.2%. Whatever biological factors do or do not combine here, this index *s* shows a variability that is certainly minimal – in the region of 0.8% – but which leaves no room for doubt. Thus, the conjecture expressed by Condorcet at the end of the 18th century has proved well-founded from an empirical point of view; and Halbwachs was not mistaken – even though he did not employ such a precise sifting process as we can here – when he observed certain variations in the male sex ratio at birth for the period he studied in the 1930s.

Rather than attempting a general history of the sciences, our investigation of the *longue durée* of the study of human sex ratio at birth has followed

---

Institute for Statistics and Economic Studies (see Halbwachs, 2005, p. 193).

Periods	1811 1815	1806 1810	1946 1950	1881 1885	1926 1930	1936 1940	1811 2000
N	930.7	923.9	860.1	934.6	748.1	596.3	886.5
M	480.7	476.0	441.4	478.5	382.1	303.8	454.6
SR	51.649%	51.521%	51.320%	51.198%	51.077%	50.944%	51.277%
T	4.65	4.62	4.30	4.67	3.74	2.98	168.44
CI95	±0.046%	±0.047%	±0.048%	±0.046%	±0.052%	±0.058%	±0.008%
Min	51.603%	51.474%	51.271%	51.152%	51.025%	50.886%	51.270%
Max	51.696%	51.567%	51.368%	51.245%	51.128%	51.002%	51.285%

N, annual mean of the number of live births for the period (in thousands).
M, annual mean of the number of male births for the period (in thousands).
SR, proportions of male births for the period ($SR = M/N$).
T, total number of births for the period (in millions).
CI95, amplitude of the 95% confidence interval ($CI95 = 1/\sqrt{T}$).
Min, minimum of the 95% confidence interval ($Min = SR - CI95$).
Max, maximum of the 95% confidence interval ($Max = SR + CI95$).

the thread of an arbitrary but crucial question: how have the form and the possible variability of the phenomenon been thought about and understood? Taking stock of this, we may doubt that human sex ratio at birth is constant. This is a question of empirical fact, not of a principle to be adopted. What is more, to accept the constancy of the sex ratio at birth would be to skip two and a half centuries of academic tensions which have affected – and still affect – the very bases of the social sciences, the way in which these sciences grasp the different forms of uncertainty inherent in the phenomena they study, and finally their relationships with other sciences – mathematical, biological or historical.

2. A SLIPPERY VARIABILITY?

Ever since Antiquity, people have been envisaging some deliberate action that might favour the birth of a boy or even of a girl: this can be traced down through various texts (Halbwachs, 1936 [2005]). In the 18th century, as we have seen, once the proportion of births of boys to births of girls became the object, the issue moved out of the sphere of praxeology, specific to action within the family, and passed into the realm of epistemology, where one or several probable causes are sought. Laplace himself (1812, 1814), in the commentaries on his analytical theory as applied to the probability of the sexes at birth, was invariably careful to reinforce this demarcation. After a century of attempts to identify particular causes, Gini's thesis (1908) hovered around this divide, finally establishing it with early-20th-century statistical methods. Thus, since then, no author has been able to risk proposing such action and expect to be taken seriously. Korherr, the SS Inspector of Statistics, was very well-informed and, in fact, obsessed by calculation; he painted the possibility of new advances in glowing colours to his boss (1942), but to no avail. Even quite recently, some less well-informed authors have hoped to provide what they view as reliable recipes; they seem to be blinded by a puzzling kind of medical or health-food proselytism – indeed, even by a eugenicist concern (Regnault, 1936; Stolkowski, 1984; Stolkowski & Choukroun, 1991). Only persistent superstitions – which Laplace answered long ago – ensure a satisfactory commercial reception for such works, which barely avoid charlatanism.

But, as we have just seen, the variability of the sex ratio at birth is an empirical fact. Nowadays, we have three scientific registers in which we can analyse it: biology, economics and even sociology. In the last arena, an attempt at this analysis was made by Halbwachs (1933, 1936). He took as his starting-point a general Durkheimian concept where the issue of sex ratio was of no importance, and according to which the collective social body changes under the influence of systems of representations. Halbwachs, following an epistemological revision of Durkheimian objectivism, rediscovered certain aspects of Condorcet's stochastic conception, according to which the collective action of humanity – or of society – would not operate directly but through the probability of its effects. However, Halbwachs also hoped to highlight a "link in the chain" – the dependency between parental age gap and the probability of the sex of the children, between socially constituted systems of representations and the biological conditions for the reproduction of social agents. However, not only was his thesis somewhat forgotten during the second half of the 20th century, but nowadays the test of empirical calculation enables us to establish that this "link in the chain" does not hold (Brian & Jaisson, 2007; Halbwachs, 2005). It is difficult to go any further along this road without constructing somewhat differently the way the three components of this attempt fit together: the probabilistic approach (a characteristic that is epistemological in nature), the consideration of logics specific to collective representation phenomena (this time, the characteristic is sociological in nature) and the consideration of physical information from the bodies of the social agents (that is, the biological fact of their existence). The issue of a sociological approach remains relevant because the majority of variations taken up or found by Halbwachs – some of which had already been indicated by Poisson (1830) – stand up to empirical tests for the periods that they could have known: notably, signs of the immediate effects of wars, and differences between urban and rural births, or between legitimate and illegitimate births, in the 19th and early 20th centuries.

Nowadays, records of large variations in the proportion of the sexes at birth from one country to another offer a different kind of convincing proof (see Table 5, p. 151–152). The first question that then arises concerns the empirical range of variability of the proportion of boys among live births. We know that it has reached more than 55% in China today; while the majority of wealthy countries present proportions of boys among newborn babies in the order of 51.15% to 51.35%. But what is the minimum observed? Although registers from late-20th-century Mexico do not appear to cover more than nine births out of ten (regardless of sex), they offer a striking picture. Out of almost 13.7 million births registered for the years 1994–1998, it can be stated with a risk of less than 5% that the proportion of boys is between 50.39% and 50.45%. Out of almost 5.6 million births in 1999 and 2000, the

Table 5: Recent sex ratios for various countries

Country	Sources	Births	Boys	Sex ratio	CI95%	Minimum	Maximum
Mexico (2001)	2a	2767610	1390066	50.23%	±0.06%	50.17%	50.29%
Ecuador (2003)	2b	178549	90710	50.80%	±0.24%	50.57%	51.04%
Guatemala (1999)	1	360759	183621	50.90%	±0.17%	50.73%	51.07%
Sri Lanka (1996)	1a	340649	173603	50.96%	±0.17%	50.79%	51.13%
Chile (2003)	1	234486	119963	51.16%	±0.21%	50.95%	51.37%
United States (2002)	4	4021726	2057979	51.17%	±0.05%	51.12%	51.22%
France (2002)	1	761630	389981	51.20%	±0.11%	51.09%	51.32%
Morocco (2001)	1	541298	277242	51.22%	±0.14%	51.08%	51.35%
Brazil (2003)	2b	2822462	1445825	51.23%	±0.06%	51.17%	51.29%
Germany (2003)	1	706721	362709	51.32%	±0.12%	51.20%	51.44%
Japan (2003)	1	1123610	576736	51.33%	±0.09%	51.23%	51.42%
Netherlands (2002)	1	202083	103734	51.33%	±0.22%	51.11%	51.55%
Russia (1994)	5	1363806	700191	51.34%	±0.09%	51.26%	51.43%
Canada (2002)	1	328802	168842	51.35%	±0.17%	51.18%	51.53%
Colombia (2003)	2	697029	357987	51.36%	±0.12%	51.24%	51.48%
Egypt (1999)	1	1693025	870195	51.40%	±0.08%	51.32%	51.48%
Thailand (2000)	2a	773009	397523	51.43%	±0.11%	51.31%	51.54%
Australia (2003)	1a	251161	129193	51.44%	±0.20%	51.24%	51.64%
Poland (2003)	1	351072	180634	51.45%	±0.17%	51.28%	51.62%

(*Continued*)

Table 5: (*Continued*)

Country	Sources	Births	Boys	Sex ratio	CI95%	Minimum	Maximum
Kazakhstan (2003)	1	247946	127610	51.47%	±0.20%	51.27%	51.67%
Malaysia (2000)	2	442502	227833	51.49%	±0.15%	51.34%	51.64%
Peru (2002)	2ab	355870	183326	51.51%	±0.17%	51.35%	51.68%
Romania (2003)	1	212459	109497	51.54%	±0.22%	51.32%	51.75%
Spain (2002)	1	418846	215995	51.57%	±0.15%	51.41%	51.72%
Venezuela (2002)	1a	492678	254969	51.75%	±0.14%	51.61%	51.89%
Philippines (2002)	1	1666773	866521	51.99%	±0.08%	51.91%	52.07%
Pakistan (2001)	3	3719694	1942845	52.23%	±0.05%	52.18%	52.28%
South Korea (2002)	1	494625	259123	52.39%	±0.14%	52.25%	52.53%
China (2000)	6	13794000	7460000	54.08%	±0.03%	54.05%	54.11%

Source: (1) UN Yearbooks for 1995 and 2003, taken from national government figures, registrations more than 90% complete; (2) *ibidem, idem*, registrations less than 90% complete; (3) *ibidem, idem*, estimates and/or regions missing; (4) *National Vital Statistics Reports* (2005); (5) Avdeev and Monnier (1996); (6) Cai & Lavely (2003); (a) the reference date is that of registration and not that of birth; (b) some indigenous South American population groups are not included (see also: unstats.un.org). In the case of Mexico, we have used figures for the year 2001, taken from the UN report for 2002, rather than those for the year 2003, which appeared the following year. This is because, firstly, the total of births for 2001 exactly matches the total published by the Mexican National Institute of Statistics (whereas for previous years there is a difference of around 0.01% to 0.03% between these two sources), and secondly, the Mexican table for the year 2003 in the last UN report seems to be incomplete when the general figures and those for the urban and rural regions are cross-referenced.

equivalent proportion is between 49.97% and 50.05%.[2] Therefore it must be acknowledged – revising Laplace's conviction – that, at the scale of large countries, variations in the sex ratio at birth may in fact very well range from 50% to 55%, and that the sociological question raised is by no means trifling.

More than 60 years ago, the economist Edmond Malinvaud observed that the level of the proportion of births according to sex varied according to family composition (Malinvaud, 1955). Similarly, beyond the neonatal period, demographers agree that factors in the social environment must be taken into account when analysing infant mortality (Masuy-Stroobant, 2002b) and differentials in premature death according to sex (Vallin, 2002). If we now take the view that sex recognition operates as a symbolic instrument (Bourdieu, 1972, 1980, 1998; Goffman, 1977; Héritier, 1996, 2002; Yacine, 1988, 2006), then the fact that forms of discrimination by sex at birth differ according to social conditions can easily be interpreted from a sociological point of view: those involved in births are not uniformly tolerant towards these explicit or implicit forms of discrimination. Research by economists on China and on India confirms this. For example, wealth indicators can be constructed for birth environments – whether these are composite indices of the level of resources in the family circle (Guilmoto, 2005) or more subtle criteria, such as the price of tea in China, which allow a distinction to be made between female wealth and male wealth (Qian, 2005). These provide a pertinent way of explaining local variations observed in the distribution of the sexes at birth. In India, better-off family circles are also more dangerous for girls, since selective abortion is within means of their resources (Guilmoto, 2005). Among Chinese peasants, although increasing income alone seems to have no effect on the level of the male sex ratio at birth, increasing relative female income increases survival rates for girls, whereas increasing relative male income decreases them (Qian, 2005).

[2] According to sources from the UN Demographic Yearbook:

Years	Sex ratio	CI95%	Births	Boys	Girls
1994–1998	50.42%	±0.03%	13 727 604	6 921 288	6 806 316
1999–2000	50.01%	±0.04%	5 566 390	2 783 687	2 782 703

NB: According to the standards applied by the UN, Mexican birth registers for these years cannot be considered at least 90% complete.

Here, reasoning at the scale of a whole species would lead to difficulties in recreating the logics that economists or sociologists apply on a case-by-case basis. This is because they are internal to the species and are based on the distinction between the two sexes – that is, on systems of representations and of resources that the two sexes more or less share, through which they think of themselves as different and consequently act differently. Here we are again at the point that we had reached at the end of Chapter 4. It is impossible to go any further, to cross from one shore to the other – from biology to the economic and social sciences (or back again) – without some risk. We would be deceiving the reader if we were to skirt around what, for at least a century, has been the distinctive feature of each of these two categories of sciences: on the one hand, the explicit or implicit comparison between species (Darwin, 1871) and on the other hand, the very fact of systems of representations (Durkheim, 1912).

The historical approach that we have taken has enabled us to highlight another difficulty. As we have already noted, the historical journey of the formation of those sciences is intimately involved with the creation of what Condorcet would have called "the technical methods" needed for these analyses. Condorcet, Laplace, Fourier and Poisson worked on the earliest formulations of the analytical calculus of probabilities. Fourier and, especially, Quetelet and those who carried on the latter's work in 19th-century European offices of statistics created procedures for gathering registrations, taking as their starting-point a simplification of this analytical calculus and, within it, limiting themselves essentially to methods based on calculations of averages. Galton, Lexis, Gini and Fisher, in their turn, broke the bounds of the simplifications established by Quetelet and adhered to by his successors. In doing so, they broadly helped to shape statistical methods that have developed since the early 20th century, which have resulted in the techniques employed today, both in the social and economic sciences and in biology (see Armatte, 1995; Armatte & Dahan, 2004; Hald, 1998; Porter, 2004; Stigler, 1986).

The elements relating to the proportions of the two sexes that were provided by the exceptional collaboration between the mathematician Fréchet and the sociologist Halbwachs in *Le Calcul des probabilités à la portée de tous* [*The Calculus of probabilities for all*] (1924) have remained completely atypical in that regard: their concern was not so much methodological (although they did initially intend it to be a technical training manual) as philosophical in nature or, to be more precise, epistemological – although the word would have been an anachronism. Fréchet and Halbwachs were simply keen to display the greatest theoretical rigour from both the mathematical and the philosophical points of view. Thus this book concerned itself with a theory of knowledge that was based on the need to use the calculus of probabilities in approaching certain phenomena. This feature is characteristic of Condorcet's metaphysics of the calculus (Brian, 1994a) and of Halbwachs' mature sociology (Brian and Jaisson, 2005b). The merely methodological use of a body of statistical techniques – whether these are processes familiar to Quetelet, processes defended by Fisher, or even an entirely different particular method developed later – does not necessarily demand an epistemological concept that is probabilistic or stochastic in nature. Physicists are familiar with this kind of distinction: for them, making use of the toolkit offered by statistics is one thing; using a quantum-based conception of physics is quite another. This comparison throws light on the precise distinction that we are seeking to make, without creating confusion between the objects and the methods of the physical, biological, economic and social sciences.

It is therefore necessary to distinguish between the often humdrum use of routine statistical techniques and the deliberate conceptual choice, which is epistemological, that would lie in taking the view that the phenomenon itself is probabilistic in nature (Condorcet and Halbwachs gave examples of this) and that therefore no one can grasp it except through estimation or by the scientific calculation of probabilities (the calculation that made Jakob Bernoulli choose the word "stochastic"). When calculating the proportions of the two sexes at birth, it is easy to compare the implications of using a statistical method, on the one hand, with those of the choice of a stochastic hypothesis, on the other.

Studies of this topic as reported in present-day publications propose *statistical modelling* of one component of the macroscopic phenomenon. The form of the model is the consequence of a choice of method. An individual birth is then considered as an established fact; the sex of the newborn baby is an initial, defined piece of data. Establishing the total of births, N, consists in obtaining a supposedly perfectly-known count of all possible cases. Among these, the number of boys, M, and the number of girls, F, do not pose any particular problem. The sex ratio, $S = M/N$, is a defined value. It is generally available for several consecutive years, S_t, or for various groups of observations, S_i. The variability that can be observed from one value to another is taken as the element of uncertainty that needs to be reduced (and therefore S can be considered, from a mathematical point of view, as a *random variable*). The notations and the terminology used then vary according to the type of model employed: "dependent variable" or "time series", for example. Let us use a time-series model S_t: this sex ratio, variable from year to year, can then be broken down into two elements: firstly, the modelled sex ratio $\hat{S}_t$ and secondly, a zero-mean "noise", ε_t; these two components are constructed as if their fluctuations were independent. The width of the "noise" is used to calculate whether the observations S_t coincide sufficiently with the model $\hat{S}_t$: thus the model is tested through comparison of the part of the variance in S_t carried by each of the two independent terms $\hat{S}_t$ and ε_t. In the end, this comparison lends itself in various ways to calculation of the probability of whether S_t conforms to $\hat{S}_t$ or not.

If it is now proposed to go on to a *stochastic reconstruction* of the phenomenon, each of the same elements will be involved in some other way. The objective becomes the analytical construction of the relationship between the basic individual piece of data and the overall assessment of the phenomenon (in other words, between the "micro" and the "macro"). There is nothing that *a priori* governs the form of the macroscopic calculation that will be obtained at the end of the reconstruction. For an individual birth, the registration of one sex or the other is the realization of a *random variable* that takes the value 1 or the value 0, with 1 for a given sex (e.g., "boy" – the choice is completely arbitrary) and 0 for the other cases, with respective probabilities s and $(1\text{-}s)$. The total number of cases, N, is supposedly fully known: according to the hypothesis, the uncertainty being analysed is that of the allocation of cases to different sexes and not that of the number under consideration. But the number of boys, M, (like the number of girls) is necessarily random: this because it is the sum of N basic random variables, each associated with an individual birth. In these conditions, the sex ratio, $S = M/N$, is the realization of a binomial random variable (which we can call Σ), whose law is known: it is properly approached through a normal distribution with expectation s and standard deviation $1/(2\sqrt{N})$ (see Appendix D). It can be inferred from this, for example, that:

$$\text{Proba}\left(|s - S| \geq \frac{1}{\sqrt{N}}\right) \leq 5\%$$

Consider, for example, the observed sex ratios for several consecutive years, S_t, which are the realizations of a series of random variables Σ_t. One could refer to a discrete stochastic process (or even a finite one).

In the same way as before, the probabilistic behaviour of S_t is sufficiently well-known:

$$S_t \rightsquigarrow \Sigma_t \approx \text{Normale}\left(s_t\ ;\ \frac{1}{2\sqrt{N_t}}\right)$$

The reconstruction of variations of S_t can be broken down between, on the one hand, an arbitrary equation of mathematical expectations of the random variables Σ_t according to time t and, on the other hand, for each year, a stochastic fluctuation around this expectation. This, for example, is within $\pm 1/\sqrt{N_t}$ at a 95% confidence level. The arbitrary nature of the choice of equation used to recreate the variations of expectations of Σ_t can then be submitted to a χ^2 test.

Let us take an example. In recent decades, articles that have appeared in the scientific literature on human sex ratio at birth have relied on *statistical modelling* methods[3]. It would be futile to review all of them, even though it is most frequently the case that each one highlights a particular factor. Ralph Catalano, Professor of Public Health at the University of California at Berkeley, has published several texts of this type, which all aim to highlight the effect of stress in reducing the level of the sex ratio at birth (Catalano, Bruckner, Gould, Eskenazi, & Anderson, 2005; Catalano & Bruckner, 2006). The authors want to mount a persuasive defence for this series of articles, but – from different starting-points – they always seem to draw the same conclusion; they provide a classic example of the type. In these studies, stress seems to be omnipotent, but it is in various concrete forms – for example, as related to the hypothetically lost generations in Swedish history (2006), or characteristic of "California after the 9/11 terrorist attacks in 2001" (2005). The time series analysis technique (Box & Jenkins, 1970) is applied to the number of boys per 100 girls.

Yet this index is not the best one for obtaining a precise measure of a random phenomenon (see Appendix D). It results in a macroscopic statistical aberration that can be discerned particularly in the case of the post-9/11 Californian study. During the last four months of 2001 (if we take the graphs in the published article), about 103 boys were born per 100 girls, which is 50.7% of births. Yet, in that year, 269,237 boys and 258,121 girls were

[3] Authors are frequently at a loss when faced with the impossibility of giving a reasonable explanation for the trends observed, and tend to diagnose "multifactorial" causes (see, for example, Grech, Vassallo-Agius, & Savona-Ventura, 2003). The technique is the same, the conclusions remain cautious, but analysis of the phenomenon gains nothing.

born in California, which is 51.05% boys (Bindra, Christensen, & Ficenec, 2004). The monthly mean of all births was 43,948. Each month, in terms of a 95% confidence interval, this is an order of magnitude of ±0.48% for the proportions of boys, or ±1.9 points around the rates of boys per 100 girls (see Appendix D). Therefore, the Autumn 2001 figures cannot be reasonably distinguished from the annual mean: there are no grounds for diagnosing a fall in the sex ratio.

That is not all: we then come to the difficulties raised by the statistical modelling itself. Using the time series method presupposes that variations in these numbers of boys per 100 girls will be calculated for specific, established figures. Yet in fact, they most likely fluctuate within a 3.8 width interval centred on the values obtained... The result is that almost all the points that would represent the level of the monthly number of boys per 100 girls from 1989 to 2002 could be placed within a constant-width horizontal band (for comparison, see Catalano et al., 2005, Figure 1, p. 1223). Thus variations in the sex ratio for California at the monthly scale are not relevant during the period under consideration.[4]

3. A STOCHASTIC RECONSTRUCTION

Mobilizing a statistical toolkit in a purely technical or methodological way thus leads to a futile attempt to create linkages between distinct elements that have settled in the different branches of sciences and are certainly all connected with the history of the problem to be tackled; but it does not lead to their being arranged into a single solid construction. So should the fragility of the "least worst" explanations accepted nowadays – with their puzzling déjà-vu and the inconsistency between them – make the scholar timid? We do not think so: the long history of the formation of several scientific disciplines around the issue of human sex ratio at birth must be taken seriously. The evidence, the phenomenon and the object should be not only deconstructed, but reconstructed again using the elements highlighted in the historical investigation, the analysis of the form of the phenomenon. A metaphor drawn from archaeology is not out of place here, in that it retains the idea of the resistance of objects. We have two or three centuries of various fairly specialized types of both material and cognitive

[4] From a strict statistical standpoint, a part of the criticism here – made from an epistemological point of view – can be expressed by showing that the calculation is not sufficiently robust for the random fluctuations observed.

sedimentation, traits of collective memory, libraries of books, scholarly techniques of all kinds. Up to now we have excavated, extricated and dusted off our solid finds. Next we should move on to some reconstructions, provided that certain artefacts fit together effectively, then ask ourselves whether these reconstructions look like something, and question the way in which we have thought of that thing. Except that here everything we need to manipulate in this way actually relates to systems of representations – and often very elaborate ones... That is why the approach we are now suggesting – on principle an epistemological one, and therefore likely to interest adherents of different disciplines, a part of whose history relates to that of the study of sex ratio: mathematicians, biologists and sociologists – finds both its framework of rational control and the terrain for its formation in recent renewals of the social sciences.[5]

We can associate a first hypothesis with the name of Laplace: when considering a child yet to be born, before its sex is recognized, its family – like scholars, whether studying this particular birth or a much larger set – can only conjecture. Scholars, and perhaps parents, can behave as if the sex of this newborn baby depended on the toss of a coin, where the parity of the two sides is not necessarily established. There is no mystery here: it is simply a matter of considering that the phenomenon is *stochastic* – i.e., it cannot be reached through reason without taking seriously the uncertainties that it entails – in the sense that Jakob Bernoulli attributed to the word: *Ars Conjectandi sive Stochastice* [the art of conjecture or the stochastic] (1713, p. 213).[6]

A second hypothesis consists of considering a set of births and accepting, following Condorcet, that the level of the proportions of births of the two sexes can be attributed to action on the part of the population concerned (Condorcet was thinking of humankind) that *affects the probability* of the birth of children of one or other of the two sexes, or, as he put it, "the means of producing male or female children [...] with a certain probability" (see Appendix A). Let us clarify that there is a combination of factors and means that can produce this effect: interventions in the strictly biological ontogenesis of the formation of the sexes, forms of selection and gender

[5] In Sections 6.3 and 6.4, we summarize Parts 3 and 4 of our book, Brian & Jaisson, 2007.

[6] See Georges Guilbaud (1952) and Norbert Meusnier (1987), who recall that, in accordance with its etymology – throwing the javelin towards a target – *stochastic* is a matter of the art of lining up the sight with the goal, and therefore means taking action in a situation that is by definition uncertain. This operates at the level of praxeology as understood by Ludwig von Mises (1881–1973).

discrimination at or before birth, devices for gendered social recognition – from under-registration of one sex or the other to accentuation of markers of sex in the case of hermaphrodite births, for example.

In these conditions, the number of registered births provides a very exact measure of this polymorphous action: it represents the number of children accepted for entry into social life and recognized on that occasion as belonging to one sex or the other. Nineteenth-century statisticians deplored the imperfections of figures compiled from birth registers, since they were seeking a positive, rigorous measure of biological births. Many researchers in the social sciences nowadays scorn figures compiled from birth registers because historical investigation can demonstrate that any immediate reading of them is inaccurate. Restricting oneself to these two attitudes would mean remaining in some strange way a victim of the positivist illusion shared by the former group and feared by the latter, neither of which has been given the means to objectify such an illusion in order to grasp the numbers of registered births in a new way.

But because the adjective *social* first of all described a way of being in society, as in Sieyès or Condorcet in particular, the idea of "social being" carries with it a certain ambiguity in this regard. For 19th-century authors, did it mean a human being with certain rights that conformed to nature or to a social contract? This being would of necessity be over the age of reason, would even have reached the age that leads to obtaining these rights. With Durkheimian sociology, a biological presupposition, sustained by the positivity of statistical registrations, replaced such legally-based descriptions. Birth, at least from the point of view of civil registration, serves as the starting-point for socialization. Since the final decades of the 20th century, changes in the early hold of medicine, biology and the law over human beings, and numerous commentaries on these changes, must suggest to the sociologist that a breach is opening in these still rudimentary Durkheimian foundations (Boltanski, 2004; Memmi, 2003). Registration therefore seems to be a particular moment that may be viewed as common to all social agents (even those who fail to be registered, because they are later faced with the consequences of this state of affairs). Thus, as we have seen, registration is the trace of the phenomenon that has to be studied. Therefore we shall consider it to be the *exact record* of the phenomenon, without having to describe it from an *a priori* biological, legal or sociological point of view. This is a third element drawn from our historical investigation.

The method clarified by Poisson, that of the binomial calculus, supplies a fourth, long-established element. If it is considered for one year and one country that, for example, all boys have the same unknown individual probability (s) of being officially registered, then the ratio of the number of boys to total births ($S = M/N$) and that unknown probability (s) are linked by a formula that expresses a second probability: the probability that their difference (S-s) is greater than a constant quantity divided by the square root of the total number of registered cases (N). One often (wrongly) talk about "the law of large numbers" when describing this result (see Appendix D).

Although the calculus poses no particular difficulty nowadays, it is important to stress that the use we make of it is not strictly technical: it now becomes a fifth element in our construction. The application of the binomial calculus to the case of the sexes at birth is not arbitrary. Fréchet and Halbwachs knew this very well when they gave an example of this type in their 1924 book. Here, mathematical formalization follows the abstraction of the phenomenon rigorously: each individual case remains governed by an intrinsic uncertainty; the binomial calculus is therefore the *ad hoc* method – from both the historical and the technical points of view – that will enable us to move from the individual scale to the scale of an aggregate. The crucial issue here is that from a mathematical point of view this modelling conforms to the stochastic hypothesis envisaged by Laplace and later developed by Poisson. At each level, from one particular birth to hundreds of millions being taken into account, the structure is coherent: no break here between "micro" and "macro".[7] Although it is true that, customarily, fragments of calculations are combined with thinking drawn from two and a half centuries of research into the proportion of the sexes at birth and into binomial distributions, this is usually done without taking the time to ensure that the combination is coherent; so our epistemological concern is to go beyond routine techniques. This means accepting that the investigation falls within a general epistemology based on probabilistic thinking, something envisaged first by Condorcet (Brian, 1994a) and then by Halbwachs (Brian & Jaisson, 2005b), each in his own way and without Halbwachs knowing Condorcet's position.

[7] Constructing objects of social sciences by explaining phenomena grasped at different levels of scale is an approach that has already been explored over some time: on examining changes of scale in social sciences, see Revel, 1996; on the possibility of constructing objects that remain coherent when the scale varies, see Brian & Alunni, 2001; finally, on the use of *ad hoc* demographic methods, see Courgeau, 2007.

The sixth piece of the jigsaw relates to empirical strategy. If we follow Halbwachs' work (1933, 1936), exceptional variations in the proportion of the two sexes at birth would be in the order of "one point" (from 51% to 52%, for example). More common variations would be measured in thousandths and not in hundredths. Yet, if we wish to restrict ourselves to confidence intervals characterized by a probability of over 95%, a variation of a hundredth in the proportion of the sexes necessitates a count of 10,000 cases, while for a variation of a thousandth we need a million cases. The strategy used in what follows will consist in increasing the number of observations, as far as administrative records allow, in such a way that the confidence intervals are smaller than the possible variations observed for the sex ratios.

Let us sum up these six initial points: (1) a heuristic hypothesis: that the phenomenon under study is stochastic; (2) a sociological hypothesis: that collective action affects the probability of properties recognized in newborn babies at the moment of their entry into social life; (3) an empirical hypothesis: that the phenomenon under study is measured by registrations; (4) an element of mathematical thinking: the formal link between the probability of one of the sexes at birth, the confidence interval around the sex ratio, and the square root of the number of cases observed; (5) the assessment that the four preceding elements are coherent and, as a consequence, that this analytical construction can rigorously reconstruct all scales of analysis, from the most singular individual case to the widest aggregates; (6) an empirical strategy: to adapt the scale of the observation (that is, the confidence intervals around the probability of one of the two sexes at birth) in order to highlight its possible variations.

With the combination of these hypotheses, the annual proportions of boys at birth – let us call them S_t – are viewed as estimates of an abstract thing, but a thing that is indispensable from the point of view of reasoning according to the calculus of probabilities: the mathematical expectation – let us call it $E(\Sigma_t)$ – of the random variable that is the sum, Σ_t, of all known individual cases of live births, which (conventionally) takes the value 1 if the child is registered as a boy and the value 0 if not.

Rejecting the constancy of the sex ratio at birth means giving up a view of this expectation as constant over time. In his publications of the 1930s, Halbwachs noted a regular secular decrease in the sex ratio at birth in France in the 19th century (S_t). In formal terms, a secular trend like this can be expressed as a linear-type model, constructed starting from the observation date t (given an arbitrary origin of the times, t_0), which can be written – choosing a notation that pays homage to the scholar who first discussed the matter most clearly:

$$\Sigma_t = s_0 + H\ (t - t_0) + \varepsilon_t$$

Therefore:

$$\mathrm{E}(\Sigma_t) = s_0 + H\ (t - t_0)$$

If we consider the annual compilations of registrations in France since the mid-19th century, it is easy to show the string of two linear models of this type, characterized by two more or less steep slopes (H), one negative and the other positive. They intersect at a breakpoint 1897 and 1898 (see Figure 3, below). This already gives us three results: a breakpoint date and two slopes. Not only does the proportion of live male births show identifiable long-term variations, but these variations also have trends that may be

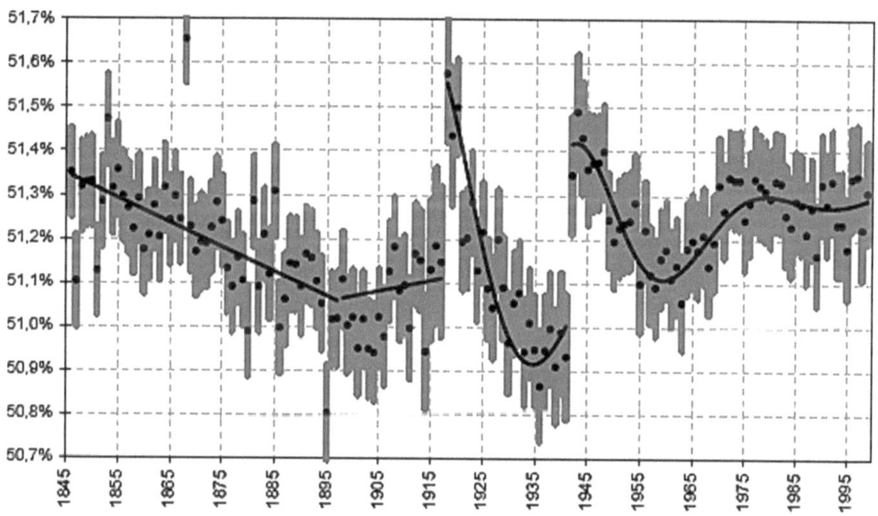

Figure 3: France. 1846–1999. Observations and model. The black dots show the annual level of the male sex ratio at birth (S_t). The grey areas are 95% confidence intervals. The black curve, apart from the two breaks, links the estimated mathematical expectations $\mathrm{E}(\Sigma_t)$ obtained using the proposed model. See its equation p. 164.

upward or downward (0.0024% a year in the 20th century and –0.0056% a year in the second half of the 19th century). The slope of the regression line of the proportions of one sex in all births therefore gives a precise index that can be calculated rigorously for a phenomenon that, for a very long time, almost always seemed to be equal to itself.

Once this first stage has been passed, we can calculate the differences (ε_t) between the known values of the sex ratio (S_t) and the values generated by these linear models, which form lines with changing slopes. Yet these differences themselves present identifiable shapes. First of all there are distinct breaks: in 1918 and 1942, the values taken by the sex ratios are markedly higher than those for preceding years, and well beyond the limits of 95% confidence intervals. Reflecting on Halbwachs' morphological reading of the stark increase in the proportion of boys in 1918 (see p. 134), bearing in mind the dates of the two world wars that France was involved in, and taking serious note of the fact that the dates of these breaks (1918 and 1942) are pure products of calculations, it is pertinent to match these morphological markers to these "major events": indeed, "some major upheaval must have taken place", as Halbwachs wrote.

Next there are regular upward and downward movements, strongly attenuated over time. These fluctuations are in fact fairly simple from a mathematical point of view, with recognizable damped oscillators. The equation then becomes:

$$\Sigma_t = s_0 + H\ (t - t_0) + E_0 e^{-\rho(t-t_0)} \cos\left(\frac{2\pi}{P}(t - t_0)\right) + \eta_t$$

The term "shock" is appropriate here not only because it expresses the extraordinary differences for these two dates, but also because it can be used in relation to initial conditions that precede damped oscillatory movements. This is neither a metaphor nor an analogy, but an interpretative schema that expresses the clear formal model.

The new residual differences (here called η_t) now prove to be very similar to random fluctuations of very limited amplitudes. The periods of the oscillations after the two breaks observed are in the order of 35 years and 44 years. Figures 3 and 4 (pp. 163 and 165) illustrate the result

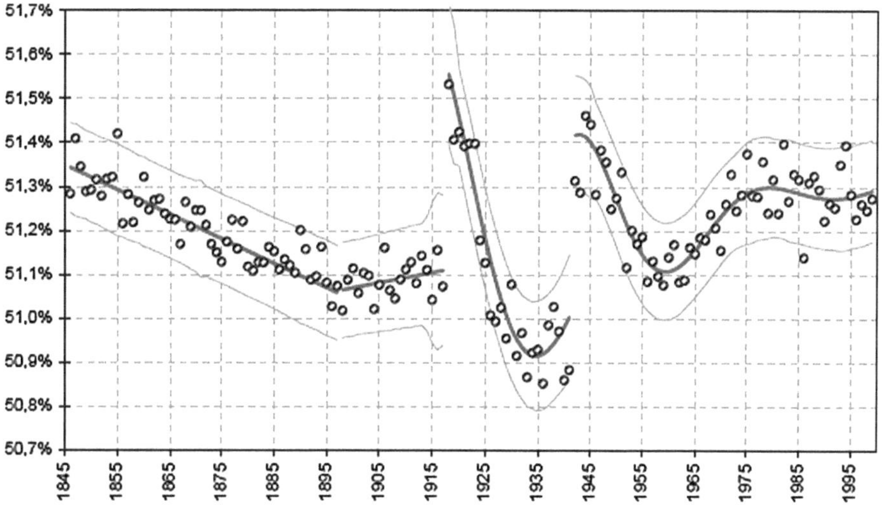

Figure 4: France. 1846–1999. Stochastic simulation. The circles show a possible realization, Σ_t, of the stochastic model. The central curve links the estimates of mathematical expectations, $E(\Sigma_t)$, obtained using the model. The curves on either side of it show the cumulative 95% confidence interval appropriate to the random residual of the model, η_t, and the known registration errors for each year.

Source: S.G.F and I.N.S.E.E. Calculations: Brian & Jaisson, 2007

obtained (the values of the parameters arising from the adjustment of the model are summed up in Table 6, p. 166). In Figure 3, the black curves represent first of all two linear sections (1846–1897 and 1898–1917), then two damped oscillators (1918–1941 and 1942–1999). The black dots show the observed levels of the annual proportions of boys at birth. The grey areas show the 95% confidence intervals around these levels, i.e., the reasonable margin of uncertainty around the observations. In only nine years out of 154 (5.8% of cases) were there proportions of boys among births that differed by more than two standard deviations from the value obtained using the model (these were 1901, 1903, 1904, 1914, 1921, 1922, 1928, 1932 and 1954).

The modelling can be tested more subtly by comparing the distribution of the residual difference between the model and the empirical observation (called above η_t) with the distribution that would produce a fluctuation for this "residual" according to a "normal" (or Laplace-Gauss) distribution. Both the calculation and Figure 5 (p. 166), show convincingly that the observations

Table 6: Parameters of the different models

$$\Sigma_t = s_0 + H(t-t_0) + E_0 e^{-\rho(t-t_0)} \cos\left(\tfrac{2\pi}{P}(t-t_0)\right) + \eta_t$$

	France 1915	France 1944	China 1981	Japan 1966
1) Absorption ρ	−0.04	−0.05	−0.16	−0.15
2) Dissipation 1/2-time T	17.3 years	13.9 years	4.33 years	4.62 years
3) Births N	387 000	623 000	18 500 000	1 360 000
4) TN	$6.7\ 10^{+6}$	$8.6\ 10^{+6}$	$80.1\ 10^{+6}$	$6.3\ 10^{+6}$
5) $T\sqrt{N}$ (to the nearest 250)	11 750	11 000	18 750	5 500
6) Pseudo-period P	44.0 years	35.0 years	5.45 years	∞
7) Period P_0	42.4 years	33.7 years	5.40 years	not applicable
8) P_0N	$16.4\ 10^{+6}$	$21.1\ 10^{+6}$	$99.9\ 10^{+6}$	not applicable
9) $P_0\sqrt{N}$ (to the nearest 250)	26 500	26 500	23 250	not applicable
10) Volume N^*	310 000	490 000	19 000 000	not applicable
11) $P_0\sqrt{N^*}$ (to the nearest 250)	23 500	23 500	23 500	not applicable
12) $\eta = \sqrt{(2\pi/P)^2 + \rho^2}$	0.148	0.186	0.164	0.150
13) Initial shock D_0	0.95%	0.45%	0.91%	0.55%
14) Tension $(\eta D_0)^2$	$2.0\ 10^{-6}$	$0.7\ 10^{-6}$	$110\ 10^{-6}$	$0.7\ 10^{-6}$
15) Mean $(\eta D_0)^2/N$	$5.1\ 10^{-12}$	$1.1\ 10^{-12}$	$6.1\ 10^{-12}$	$0.5\ 10^{-12}$

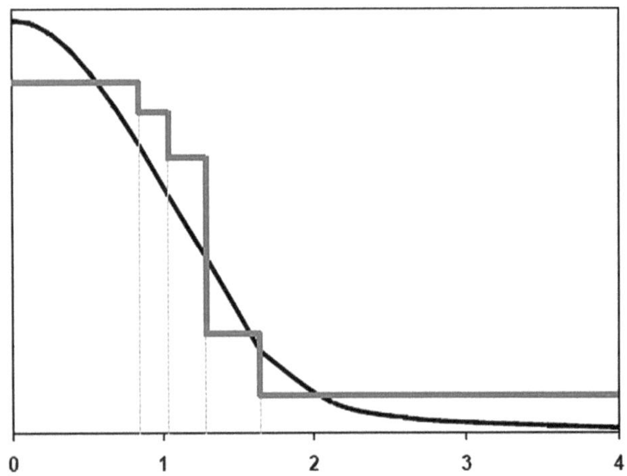

Figure 5: France. 20th century. Distribution of the difference between the model and the observations. (The horizontal axis is graduated by number of standard deviations.)

and the model fit together satisfactorily. However, the distribution of the residual difference in the model offers greater variability than the standard of the theoretical distribution. A statistical test of the comparison of these two distributions leads to a probability of rejection of their identity hypothesis in the order of 8% to 9%.[8]

Thus, when we estimate, for each year, the number of births by sex at the scale of the whole country, using a model that conforms to the uncertainty *of each individual case*, we are, as far as France throughout the 20th century is concerned, limiting our risk to only one chance in twelve. We will therefore restrict ourselves to just ten parameters and one uncertainty reduced to a simple expression (the one that appears in Figure 5), rather than using registration figures that would presume a systematic registration of 74.4 million births from 1901 to 2000. Another way of judging the coherence of the construction of the model and of the regularity of its stochastic fluctuations may be to produce

[8] Comparison for the 20th century (1901–2000) between the deviation distribution $|\eta_t|$ and the deviation from the normal distribution N(0;1).

Division into five intervals (basis of Figure 5, p. 166)		Theoretical distribution N(0;1)	Empirical distribution $\lvert\eta_t\rvert$
First six deciles	[0%–60%[	60	51
Seventh decile	[60%–70%[	10	12
Eighth decile	[70%–80%[	10	13
Ninth decile	[80%–90%[	10	7
Last decile	[90%–100%]	10	17
Total		100	100

χ^2 criterion: 8.45 for 4 degrees of freedom. Probability of rejection: 7.64%.

Division into deciles		Theoretical distribution N(0;1)	Empirical distribution $\lvert\eta_t\rvert$
First decile	[0%–10%[	10	11
Second decile	[10%–20%[	10	5
Third decile	[20%–30%[	10	13
Fourth decile	[30%–40%[	10	4
Fifth decile	[40%–50%[	10	7
Sixth decile	[50%–60%[	10	11
Seventh decile	[60%–70%[	10	12
Eighth decile	[70%–80%[	10	13
Ninth decile	[80%–90%[	10	7
Last decile	[90%–100%]	10	17
Total		100	100

χ^2 criterion: 15.2 for 9 degrees of freedom. Probability of rejection: 8.56%.

reconstructions of annual sex ratios using simulations according to these few parameters. Figure 4, p. 165, gives an instance of this, produced according to the equivalent of 124.8 million simulations of random binary trials distributed across 154 virtual years numbered from 1846 to 1999, and according to the numbers of "annual" trials equal to the total births observed for those years.

But a printed figure does not provide a suitable way of reconstructing fluctuations in probability! Here we are touching on a material condition of an intellectual task of the first importance, a condition which governs the undoubted difficulty that may be experienced when faced with a stochastic conceptualization. In order to obtain a more rigorous system of representation, it would be necessary, in Figure 3, p. 163, for example, for the black dots that mark the observations to fluctuate randomly: this is because each position is known only in probability, as the shaded representation of the confidence intervals would suggest. Also in Figure 3, only the curve of the model formed from the two line segments and the two damped oscillators satisfactorily plots the object that needs to be represented, i.e., $E(\Sigma_t)$. Similarly, though for slightly different reasons, in Figure 4, p. 165, each small circle – a marker of the annual total of individual simulations conforming to the established model – should fluctuate in the same way as the residual difference (η_t), something shown by the band plotted for the differences from roughly two standard deviations around the model.

In the general equation of the model:

$$\Sigma_t = s_0 + H(t - t_0) + E_0 e^{-\rho(t-t_0)} \cos\left(\frac{2\pi}{P}(t - t_0)\right) + \eta_t$$

E_0 appears as the deviation of the sex ratio at the initial moment of the shock and s_0 as the value of this index that would have been observed if the shock had not taken place. We have already commented on the interpretation of H: it is the long-term slope that characterizes the secular trend of the sex ratio at birth, observed by Halbwachs. P and ρ are respectively what are described in other modellings of this type as a pseudo-period (that is, the rate of oscillations of the phenomenon) and a damping coefficient of the shock (that is, the index of the speed at which it dissipates). It will be observed that these pseudo-periods are fairly comparable for the two shocks highlighted and that they are in the region of 40 years.

But a formal model brings not only economy of means when moving from several million observations to some ten parameters. Its very morphology can offer new conclusions. Thus, for example, it explains the movements of sex ratios at birth that can be discerned on the occasion of each identified shock but had remained unnoticed up to then. This is because each stark increase over a certain time is followed by a reduction that is almost as marked and lasts for a comparable length of time. This is not just a return to the previous situation, but actually a reaction. These were the years "1928–1941" and "1955–1965". Therefore this is clearly a *dynamic* in the distribution of the sexes at birth, not just – as demographers in Europe have merely repeated after Halbwachs – exceptional rises peculiar to post-war periods.

4. THE DYNAMIC OF HUMAN SEX RATIO AT BIRTH

The variation in the distribution of the sexes among declared live births in China during the three final decades of the 20th century lends itself to entirely similar modelling (Brian & Jaisson, 2007). Figure 6, below on p. 169, is the equivalent for China of Figure 3, above, for France. It would be futile to try to test the validity of the model beyond this simple graphic reconstruction, to the point that we have done in the French case. This is because Chinese registrations present inaccuracies that are well-known to specialists (Zhang, 2004), which reduce this calculation to a mere skeleton of a model. However, one conclusion does emerge: a model characterized by linear trends over many decades and

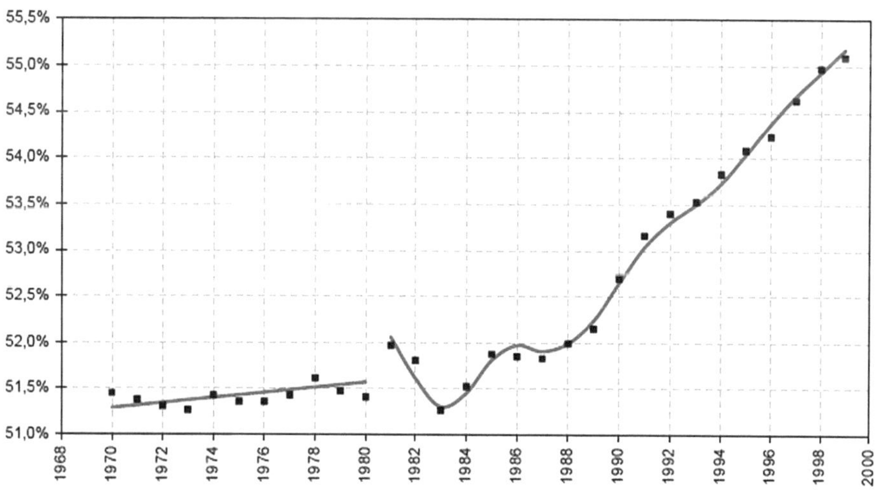

Figure 6: China. 1970–1999. Model and observations. The dots show the observations (the 95% confidence interval is here indicated by the rough thickness of the dots). The curve links the values generated by the model.

then a moment of shock followed by a damped oscillation, constructed on a few parameters and almost 610.3 million basic random simulations, follows the major lines of variations in the sex ratio at birth across three decades. Structurally, this model is the same as the one that emerges from the French case; yet China is regarded by specialists as very different. In fact, only the intensities differ: in the 1970s, the rise in the proportion of male births in China (0.0281% a year) was almost 12 times greater than in France (0.0024% a year). During the 1990s, the Chinese increase (0.2819%) was almost 120 times larger than in France.

Although all the parameters of the general level (s_0) and the secular trend (H), or even the choice of a time point (t_0), seem to agree with a certain intuition about the phenomenon, the addition of damped oscillators (ρ and P_0) may seem more artificial. The case of Japan leads us to think not. Coherent administrative sources are available for this subject, covering almost 125 years. At first glance, Figure 7, below, suggests that there is little analogy

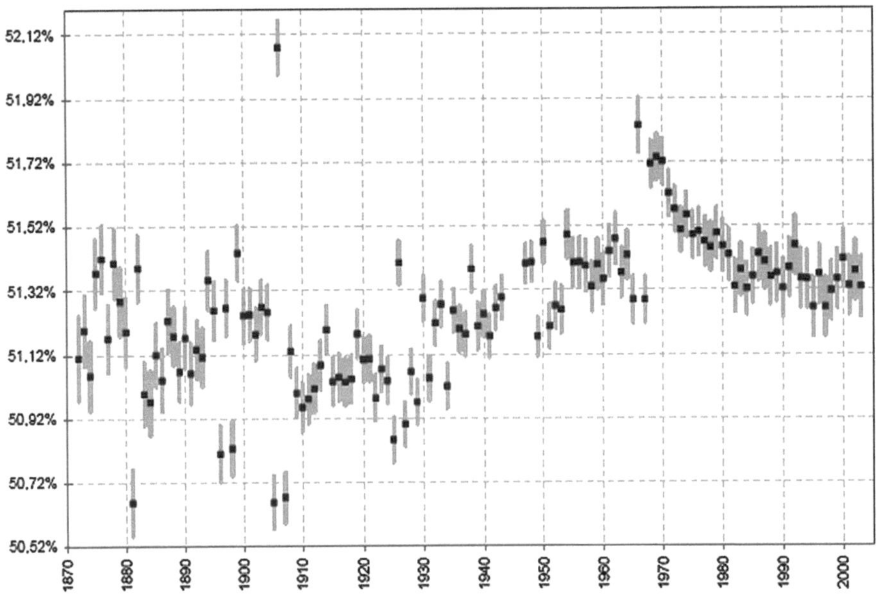

Figure 7: Japan. Annual sex ratios from 1872 to 2003. The dots show the annual observations of the proportion of boys among live births according to Japanese registrations; the vertical bars show, for each year, the 95% confidence intervals.

Source: Japan Statistical Association, *Historical Statistics of Japan, 1987–1988*, 1st ed. (Statistics Bureau, Management and Coordination Agency), Table 2–21 "Live Births by Sex and Sex Ratio of Live Birth (1872–2002)"

with the French and Chinese cases. Apart from the year 1906, a single peak comparable to those of the world wars in France appears in the mid-1960s. Then, after some decades, the proportion of boys among registered live births settles at around 51.35%. There is no oscillation shown on this Figure, only an unexpected parallel between 1905–1906–1907 and 1965–1966–1967 (see Figure 8, below). Sixty years apart, the two central years each indicate a peak in the proportions of boys registered among live births. Yet earlier and later years present a marked deficit for the same index. Confidence intervals around the values observed justify this conclusion, leaving no room for doubt. These two exceptional years, 1906 and 1966, have already long been highlighted in the literature as “fire horse years” (*Hinoe-Uma*) – years in the traditional calendar, marked by a belief that they are ill-fated for girls. Every 60 years, girls born under this sign are said to be rebellious and therefore to represent a potential danger to their future husbands (see *inter alia* Houston, 2003). In other words, the parents of a child to be born under the sign of *Hinoe-Uma* may fear these bad auspices, or even their effects in the minds of those boys or parents-in-law that any daughter they might have would later encounter.

As a result, several analysts have been able to record that parents and doctors sometimes saw fit, around 1966, to manipulate the dates on which births of girls were registered (Kaku, 1972; Usui, Houhami, & Kaneko, 1976). Confirmation can be found of the effects of these strategies specific

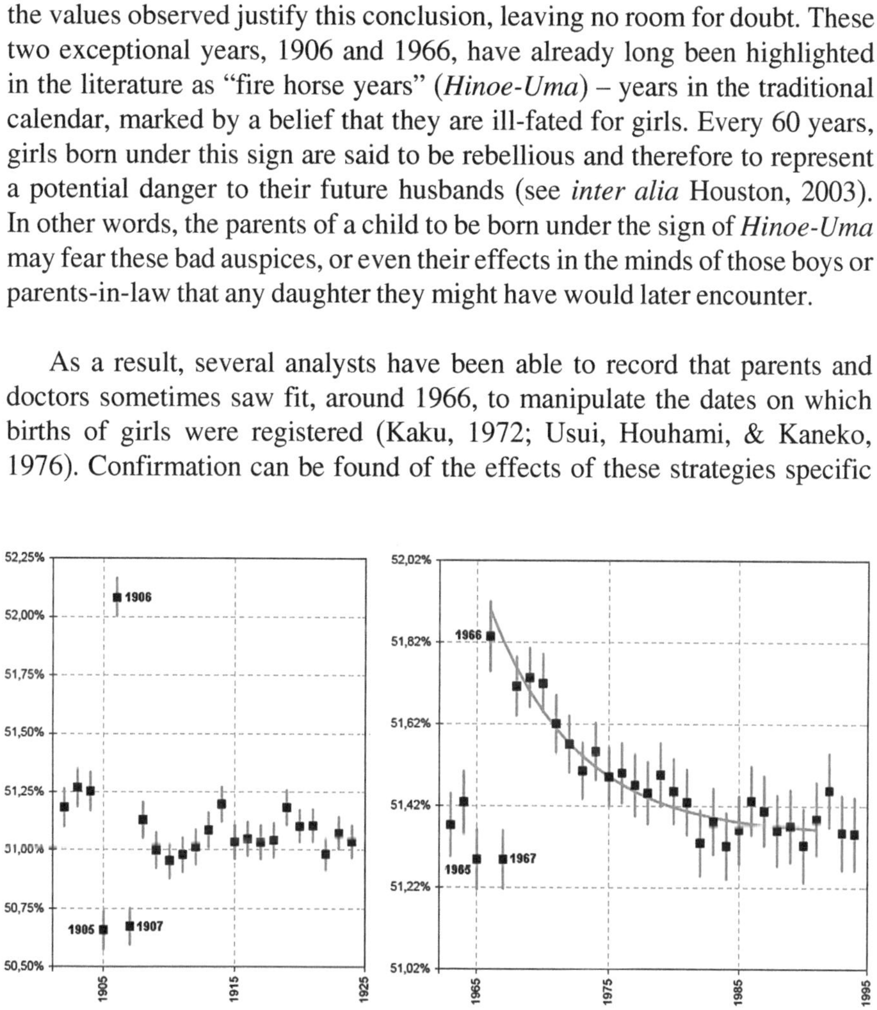

Figure 8: Japan. Annual sex ratios around 1906 and 1966. (Key as Figure 7; on the right, the grey curve indicates a model of damping without oscillation that conforms to the observations.)

to the child's immediate environment, by checking the time of birth and the date of registration of the birth, comparing the monthly compilations of birth registers for both sexes combined from Summer 1965 to Autumn 1967 (Brian & Jaisson, 2007). Looking this time at three consecutive years, from 1965 to 1967, the balance of the sexes in these annual movements is an excess of almost 1,900 boys (or a deficit of 1,900 girls), which is 0.037% of births in those three years. But for 1905 to 1907, the same calculation shows an excess of almost 11,600 girls (or a deficit of the same number of boys), which is 0.26% of births for these three years.[9] Thus, the same "fire horse years" phenomenon finds expression, essentially of variations in the sex ratio, in negotiating the moment of the newborn baby's birth in his or her environment, or even – in the case of a girl – the conditions of her official recognition. This negotiation consisted of manipulating as far as possible – regardless of sex – the date of the physical birth, or in ensuring – for girls – that registration did not bring any mark of ignominy. At the margin of this dual logic, a change in the level of sex ratio can be

[9] These elements come from the calculations below, which compare the distributions of observed live births across 1905–1907 and 1965–1967 with those that would have produced the proportions of the two sexes as measured in 1900–1904 and 1960–1964 (sources as for Figure 7). In this phenomenon, the size of the fall in births in 1906 and 1966, confirmed here, and the extent of avoidance strategies for those years makes it somewhat puzzling when attempts are made to analyse more subtle factors – for example, in the study by Rohlfs, Reed and Yamada (2005).

Annual observations:

Years	1905	1906	1907	1965	1966	1967
Annual observations of births	1 452 770	1 394 295	1 614 472	1 823 697	1 360 974	1 935 647
Observed sex ratios	50.66%	52.08%	50.67%	51.29%	51.84%	51.29%
Observed excesses of boys M–F [1]	19 126	58 015	21 756	47 035	49 952	49 909
Hypothetical sex ratios*	51.24%	51.24%	51.24%	51.42%	51.42%	51.42%
Hypothetical excesses of boys M'–F' [2]	35 976	34 528	39 980	51 640	38 537	54 810
Comparison [1] – [2]	−16 850	23 487	−18 224	−4 605	11 415	−4 901
For 3 years	1905–1907 (N = 4 461 537)			1965–1967 (N = 5 120 318)		
Results	+11 586 **girls** (i.e. +0.260%)			+1 909 **boys** (i.e. 0.037%)		

*Hypothetical sex ratios for 1905, 1906 and 1907 are calculated on the average observations of 1900–1904; homologous hypothetical sex ratios for 1965, 1966, and 1967, respectively on the average observations of 1960–1964.

seen for the two periods, which remains difficult to describe at this point in the analysis and which did not operate solely to the detriment of girls.

For the years after the *Hinoe-Uma* years, it is not possible to discern periodic effects on male sex ratios, but only a sort of damping in the single case of 1966 (see Figure 8, p. 171). Yet from a mathematical point of view, this simple damping again conforms to the general model that emerges in the Chinese and French cases: indeed, we can talk about "critical aperiodic damping" in the case where the period P_0 may be viewed as infinite. (For the other parameters of the year 1966 in Japan, see Table 6, p. 166). Thus, in this case, which is governed by a cultural trait specific to the society under consideration and consequently internalized by the parents involved, we observe either, in 1906, when it was probably easier to manipulate registrations, a very rapid re-establishing of the level of the sex ratio at birth or, in 1966, a slower re-establishing (ρ), without discernible periodicity. By comparison, the periodicity of the oscillations in the French and Chinese cases (P_0) seems to be the distinctive feature of overall responses at the scale of these two other countries to two external causes (or at least viewed as external by parents): particularly bloody wars or a very strong political constraint on fertility. These responses to external constraints demonstrate tensions that result from actions and reactions, and these remain to be analysed here (Brian & Jaisson, 2007).

In the different cases modelled (Japan in 1966, China after 1978, France during the two world wars; see Table 6, p. 166) a single date t_0 in all the terms of the equation governs the values of s_0 and E_0: the three parameters in fact offer only two latitudes (or two degrees of freedom). E_0 and s_0 are easy to interpret: $s_0 + E_0$ is the expectation of the sex ratio at the date $t = t_0$. Thus t_0 appears to be the start date of the "demographic shock". In the French and Chinese cases, these abstract dates of "shocks" are "mid-1980", "1915" and "1944" – not at all the dates recognized by legitimate collective memory: "1978", "1914", and "1939" or "1940". It is also necessary, in each case, to consider a second date: that of the first manifestations of the effects of demographic shock. This time these are respectively "1981", "1918" and "1942" – which are no more identifiable with the years viewed as legitimate markers of the events than the previous dates... What does this mean? The phenomenon under study has no reason to be directly affected by a political or military event, whether immediately or after an assumed biological time period, such as that of human gestation. Presupposing this would require a causality that would be the product of a simplistic mechanism. On the contrary, the fairly long delay between the events that seem to be at work and the date that they become manifest as a demographic shock is the proof that each event under consideration (or each period of events) sets in train a critical conjunction of circumstances, assimilated slowly by the social structure – and it is this assimilation that has an effect on the probability of births according to sex (that is, the effect envisaged by Condorcet). This effect on the sex ratio at birth is therefore mediated by a social process which can last anything from some months to some years. There is a delay in the response of the social body between the event that the collective memory retains and the demographic shock that results from it. It operates in a precise locus, the object here of the empirical study: the integration of newborn babies into the social body through a system of socially qualified representations. The numerical objectivization of collective phenomena claimed by Durkheim (1897), of which Halbwachs made a critical exploration and outlined a revised version (1936), is captured here with rigour: but now it is based on the stochastic nature of the phenomenon.

The distinctive property of a calculus is that it provides comparisons. Table 6, above p. 166, sums up the parameters of the four damped oscillator models that we have been able to construct. In the case of Japan, it will be remembered that the oscillatory dimension of the phenomenon was not apparent. Therefore, it is only the speed of dissipation of the initial demographic shock that lends itself to a comparison across all four cases. Its coefficient is the parameter ρ (see Line 1 in Table 6). Once expressed as dissipation time of half-width of the initial shock (T in Line 2), ρ is easy to interpret. In France, 17 years after 1915, i.e., in 1932, this level of dissipation of half-width of the demographic shock caused by the First World War had not yet been passed; 14 years after 1944, i.e., in 1958, its Second World War counterpart had hardly been reached. In China, the equivalent half dissipation was complete by 1985. Similarly, in Japan, four-and-a-half years after the "fire horse year" of 1966, the width of the dynamic of the sex ratio at birth had decreased by half. The parameter ρ calibrates the duration of the phenomenon and therefore provides an empirical measure of a specific social time.

Is this duration linked to the size of the population concerned, which can be estimated from the number of births at the time of the demographic shock (see Line 3)? At first glance, two types of dependencies can be envisaged: the duration could be more or less inversely proportional to the number of births, if the important thing was the strict aggregation of cases; or more or less inversely proportional to the square root of the same number, if it was important not as such but stochastically – that is, at the scale of the combination of individual variabilities of the phenomenon, whose measure is its standard deviation, and which is like $\sqrt{N}$. Neither of these two trails leads to satisfactory results (see Lines 4 and 5).

The period P_0 of non-damped oscillations[10], in the French and Chinese cases, offers another criterion of comparison. It is no longer a matter of the dissipation time of the dynamic phenomenon, but of its own distinctive rate. The five-and-a half years for China and the 30 to 40 years for France, shown here, are characteristics of macroscopic social structures whose dynamic responses can be compared (see Line 7). It will be noted immediately that the variation in these values is not simply linked to the first factor that might come to mind: the length of a generation is fairly close in the two countries. But can this period of oscillation be compared with the number of cases concerned? Although an inverse proportionality, which would signal a direct effect of the number of births, yields nothing (Line 8), it is possible to show a fairly proportional dependence with the square root of this number – and therefore with the combined individual variability of the phenomenon (Line 9). The index used to highlight the scale of the phenomenon is in fact fairly approximated: the number of births in China is not rigorously known, and in none of the countries involved does the phenomenon under study have an even effect across the whole population. Thus we can simply state that, for the three models constructed in the French and Chinese cases, it is reasonable to conjecture that the periodicity of the phenomenon is linked to the scope of the combined individual variabilities that make it up. The social rythm P_0 appears to be empirically linked to a morphological dimension of the extent of the stochastic phenomenon.

From a strictly numerical point of view, this is to say that the period is inversely proportional to the square of the number of cases concerned. Starting from estimates of this number (N*, on Line 10), we then obtain (Line 11):

$$P_0 \approx 23\ 500/\sqrt{N*}$$

From a sociological point of view, this conclusion – which nothing had led us to expect – means that, firstly, the individual uncertainty specific to each birth and, secondly, the scale of the global phenomena under consideration

[10] As usual, the period P_0 is deduced from the non-damped oscillator which would correspond to a damped oscillator of pseudo-period P and of damping coefficient ρ by this formula:

$$\left(\frac{2\pi}{P_0}\right)^2 = \left(\frac{2\pi}{P}\right)^2 + \rho^2$$

together represent characteristics of the social structure that governs the periodicity of the dynamic of phenomena under scrutiny. That it may take the most diverse forms possible is not the point: drawing up a statistical balance sheet of the numbers of births of boys and girls does not enable us to describe the various modalities that operate at the scale of analysis of the countries under study. We have chosen to call the coherent phenomenon summed up in this way "sexism at first glance" – that is, the effect of tensions peculiar to the identification and distinction of the sexes up to the time of the official registration of the birth (Brian & Jaisson, 2007). From an empirical point of view it is remarkable that the order of magnitude in $1/\sqrt{N}$ is no longer just the measure of the macroscopic variability of sex at birth (according to the principle of binomial approximation), but also the characteristic feature of the relationship between the scale of the phenomenon and its proper social tempo. At this point, the way opened by Condorcet and then by Halbwachs, that of stochastic analysis of social phenomena, leads to an examination of empirical reality that goes beyond a simple choice of method, which can always be suspected of being arbitrary. The consistency of the social fact – here comprising not just its level, but also its tempo – lies not in the number itself but in the way in which the uncertainty particular to instances of the phenomenon is aggregated into a number. To express this using two terms often employed by social scientists – it is through probability that the micro-level and macro-level are linked together.

CONCLUSION

"The decisive action of reason is almost always confused with monotonous recourse to the certainties of memory."
Gaston Bachelard (1936).

Since the 18th century, one phenomenon – the degree of regularity of the proportion of the sexes at birth – has contributed elements of support to various developments such as the calculus of probabilities, administrative statistics, the moral and social sciences, the statistics of variability, post-Darwinian biology and even Durkheimian sociology. The staying power of the object has lent itself to this. But, as Edgeworth indicated as long ago as 1892, its avatars in the various sciences do not coincide. We should add that, over three centuries, they have often crossed paths, sometimes keeping a fair trade going, but also laying themselves open to other objects being smuggled in. In this book, we have attempted to explore the phenomenon itself through a history of the way it has been understood by scholars – a history that is a reconstruction of the forms taken by the rational control of technical and conceptual operations, as faithful as historiography will allow us to be in the concrete, contextual conditions of long-standing objectivizations, and not governed by a teleological view of discovery or of science.

As we come to the end of what might therefore be described as an epistemological and historical deconstruction of human sex ratio at birth, the various pathways that we have followed through what seems in retrospect to be a history of the relationships between certain areas of mathematics, biological sciences and social sciences have given us a panoply of means that we have finally been able to recombine effectively with new reconstructions in mind. The special scholarly activity of analysing something in order to reshape it is certainly nothing new! Even before Galileo, Jupiter was known, optical lenses and telescopes were produced,

E. Brian and M. Jaisson, The Descent of Human Sex Ratio at Birth, 177–185.

sidereal movements were observed, variants of the world system had been developed... But it was a particular strict combination of these pieces of knowledge, these forms of expertise and these objects that allowed him to identify the satellites around Jupiter and, through this empirical observation, to substantiate Copernican theory. The new elements that we would like to highlight lie elsewhere. Running counter to routine narratives that skim over three centuries of research and theories of science yet care little for the conditions of its historicity, we have attempted to take a serious look at the complexity of historical times and of the actual loci of scientific work. We have highlighted them using analytical instruments drawn from the social sciences and from the history of sciences. Our close deconstruction from the standpoint of the current state of critical thinking on conditions of scientific knowledge and our concluding outline for a well-founded possibility of reconstruction therefore mean not only renewing our links with the oldest traditions of scholarly thinking, whose wellsprings undoubtedly remain hidden beneath the secrets of the scholar's craft, but also proposing a form of "well-tempered" reflexivity in today's work of objectivization. In this instance, current research on the distribution of the sexes at birth mobilizes many fragments which cannot be combined in any relevant way without thinking deeply about the coherence of the objects, the instruments – in this case, statistical ones – and the theoretical systems. Nothing would be worse than to let oneself be carried along by these routines which, for example, could lead one to use a given index, a particular statistical test or a given microeconomic model, without allowing one to foresee that these techniques may blur the phenomenon rather than providing tools for its analysis.

At least three conditions favour this movement from a series of critical deconstructions towards an attempted reconstruction. In the first place, the position that we adopt is not one that overarches the other sciences in epistemological terms – the kind of position that, in the past, would have been defended from either a normative or a critical point of view. Since the early 20th century, there have been numerous interactions between philosophy, the mathematical and natural sciences and the social sciences. As a consequence of that century, these interactions have, in one way or another, given thoughtful specialists long collective experience of intense, always somewhat fraught cross-disciplinary discussions.[1] In starting from an analysis of the history of these tensions, we take that for granted. A second

[1] Published in French since 1900, the *Revue de synthèse* has helped to animate and record these transactions, transfers and tensions. Its archive collection is available online from the National Library of France (gallica.bnf.fr). Several recent issues have been directed

factor arises from the considerable change that has occurred in the state of historical knowledge about the forms that used to be taken by scholarly research, about its most concrete conditions and about the ways in which works have circulated between the sciences and between languages. Thus it is much easier nowadays than it was 30 years ago to come straight to the point of the tensions between mathematicians of the probable in the late 18th century, to follow the pathways of Charles Darwin's writing, or to track the inaccuracies that accompanied the spread across Europe of a particular figure or a remarkable conclusion. During the same period, the pre-disciplinary history of the social sciences (Heilbron, 1990) developed significantly, providing as food for thought some elements of a fairly balanced treatment of the different disciplines concerned, before the major turn institutionalizing them in the universities in the late 19th century. Finally – and this last factor seems to us no less decisive than the others, although it is perhaps not as much discussed – the concrete conditions of the work of empirical objectivization have profoundly changed (Lepenies, 1985, 2003). Süssmilch and Condorcet could do no more than recommend that numerical tables should be drawn up. Fourier's and Quetelet's priority was the organization of administration so that counting would be carried out in a satisfactory manner. The only starting-points available to later authors were these very slightly improved registers, and figures compiled by their contemporaries who were administrative statisticians. The latter spent most of their time producing figures, using a form of administrative organization that was impressive but still imperfect. They published tables that they wanted to be viewed as definitive. Darwin, Durkheim, Lexis and Gini in turn wore themselves out poring over this material, and sometimes criticizing it. They managed to free themselves from dependency on it only through exceptional conceptual efforts. Max Weber and Maurice Halbwachs, each time they sought to explore the phenomena they were studying by any routes other than citation of extracts from collections of statistics, spent hundreds of hours going through sources in great detail, filling in and classifying index cards, and then extracting a few convincing figures from them.

Yet nowadays, if we are willing to mobilize the documentation and apply adequate processing methods to it (although it is true that this can frequently be fairly unrewarding), we have within easy reach most of the material compiled over three centuries and the mass of commentaries on

---

at promoting thinking about this long collective experience: "*Henri Berr et la culture du XXe siècle*", No. 1–2, 1996; "*Éléments d'histoire des sciences sociales*", No 4, 1997; "*Actualité et épistémologie*", No. 1, 1998; "*Sciences et philosophie au XXe siècle. L'École de Zürich et le programme surrationaliste*", No 2, 2005.

it, as well as the panoply of technical methods perfected since then, which we can use to analyse it. Current forms of academic work, the role that documentation by electronic media can play in it, and the powerful means of processing and calculation offered by personal computers allow us – notably in the social sciences – to combine, check and renew our efforts towards elements that were previously, in earlier forms of the division of labour of objectivization, dispersed across various sites. Thus a new form of academic division of labour is being outlined. It gives the scholar, at the scale of the practical conditions in which his thinking takes place, a laboratory for the "decisive action of reason" (Bachelard, 1936),[2] unlike the routines of memory associated with older divisions of labour. Figures, methods and conclusions that used to be adopted as if we had to make the best of them, can now be put to the test; the necessary historicity of their formation and their circulation can be measured; and finally we can use these early results as foundations of new terrains for empirical research.

Let us consider that only humankind has attempted to be systematic in counting itself for a few centuries. But even in this case, which is so favourable, it would be futile to expect the same methodological qualities from indices obtained by such different routes as conjectures about conception, for the primary sex ratio; empirical compilations of birth records, for the secondary sex ratio; or even censuses and their demographic extrapolations, for the tertiary sex ratio. Therefore, human sex ratios at birth offer a singular empirical case: an index which has incorporated three centuries of consistent collective efforts to establish it. Yet it is fairly easy to highlight the gap between the ideal of such a ratio and the concrete conditions for its establishment from one country to another, both nowadays and in the past, or even from one era to another. Therefore it would be futile to limit ourselves only to this ideal – and equally futile to settle for the disappointment that this suggests. We need to take as established fact – and therefore as a thing, as Durkheim would add – records made, possible compilations and calculations, imperfections in procedures, their variations, and the over-abstraction that the calculuses presuppose. The persistent divergences between the sciences, indicated by Edgeworth, no longer seem to be the tokens of a conflict of idealized faculties, but to be the indicators of a

[2] This phrase, like the quotation in the chapter heading, comes from Gaston Bachelard's article, "Le Surrationalisme" (1936). Various elements extending this position are indicated in these issues of the *Revue de synthèse*: "*Pensée des sciences*", No 1, 1999; "*Objets d'échelles*", No 1, 2001; "*Sciences et philosophie au XXe siècle. L'École de Zürich et le programme surrationaliste*", No 2, 2005.

new empirical topic to be grasped at the limit of the capacities of humankind to objectivize itself.

As soon as this perspective is set, it is possible, as we have noted, to perceive a recurring theme in the overall understanding of the phenomenon among scholars who have commented on it. Those have sought to show its dynamic have stressed its *oscillatory nature around a near-equilibrium of the two sexes*, with this feature being justified differently each time. Nowadays, specialists have in mind the wording of the principle that Fisher considered he had proved. Historians of biological doctrines are able to find prototypical formulations of this "principle" in Darwin's body of work and in Düsing. Yet we have seen that Darwin, Düsing and Fisher did not envisage the phenomenon in the same way. Each viewed this schema of a trend towards balance in a different conceptual framework: a complex combination of natural selection and sexual selection in Darwin; a re-establishing of physiological balance for Düsing; and the action of natural selection on parental expenditure for Fisher. Other instances can also be found. The cycles highlighted by Halbwachs are an example of this, although admittedly there is no idea here of underlying adjustment in the end. And very much earlier, Condorcet, in a manuscript that remained long unpublished, indicated a moral mechanism peculiar to humankind, whose effect would be the re-establishing of a balance between the sexes àt birth. In other words, each time, the same morphology of the phenomenon is expressed and, each time, this expression proceeds from a different conceptual framework specific to a particular moment in the history of relations between natural sciences and moral sciences. Similarly, when these different formulations, and the proofs that have been advanced in their support, are compared, none of them seems completely satisfactory.

A second recurring theme is worthy of attention. An inference drawn from a ratio presupposes a more or less explicit and more or less controlled conception of the *variability of the measured phenomenon*; and the diversity of these concepts seems to have governed the main differences between the academic constructions witnessed over three centuries. Is it a physical variability, an intrinsic uncertainty, measurement errors, or even errors in approximating? From a strictly technical point of view, this ceased to be very important after Quetelet, who – undoubtedly faithful to Fourier – set the seal on a powerful analogy between physical variability and error dispersion, and in doing so led to a normalization in the treatment of both, based on the hypothesis of a Laplace-Gauss distribution. It really must be admitted that the teaching of statistical methods nowadays remains faithful to Quetelet, even though it is most often ignorant of their origin: what does the nature of the dispersion matter, provided we have the central trend? Yet we must also acknowledge that the analysis of the phenomenon and any possible resulting reconstruction depend on how variability, uncertainty, errors and approximations are constituted in their empirical and technical operations.

For example, a line can be traced from Laplace to Fréchet, via Poisson, Borel and Gini, among others. This is the line of a strictly formal conception of the variability of the sexes and of the sex ratio, in different accounts of the mathematical theory of the calculus of probabilities. From this strictly formal point of view, there is little *a priori* importance in causes, errors or approximations. We merely need to note that there are two types of registration of a birth (the two sexes) and that it is impossible to allocate *a priori* a future birth to one or the other of these. Once this uncertainty about each case, which is supposed to be intrinsic and consistent, has accumulated in a certain number of observations, it allows us to infer the probability of one or the other of the two sexes for a future birth with some rigorously established degree of certainty. For each elementary calculation, the variability in frequency of one sex at birth then fluctuates in a specific way, and its measure can never be taken as fixed: it is always known to be bounded within an interval (a limited one, of course), and only in probability. This *probabilistic* schema (in the sense that it results from the mathematics of the probable) consists of accepting a necessary chance and of drawing mathematical conclusions from it. So in no case is there any question of removing the purely epistemic variability of the phenomenon. What is more, this explains its great regularity; and Borel was able to relate the empirical imperfections occasionally observed back to an impoverished physics.

Another line starts from Fourier and Babbage, then runs via Quetelet and on towards Lexis and Halbwachs. In this line, the variability of the phenomenon is no longer solely epistemic in nature, but physical. This other point of view makes it important to improve recording and measures and to perfect the treatment of errors. This is the pathway of *descriptive statistics*, which has most often appeared to be a lower priority for mathematicians of the probable. Although in the late 18th and early 19th centuries, poverty of sources made Condorcet, Laplace and Fourier promoters of what may seem to us to be a combination of the probabalistic schema and empirical statistics, the success of Quetelet's simplifications, the development of offices of administrative statistics and the proliferation of statistical publications gave *la statistique* a great deal of autonomy from the grip of mathematicians in the main Continental European countries involved. So much so that this corpus of statistics served as empirical material for the new specialisms that appeared in this era, sociology and demography. In general, for this descriptive statistics, any variation is better than none, insofar as the sources are solid and the differences measurable, since the epistemic uncertainty that the calculation entails is not taken into consideration.

A third line starts from Darwin, goes via Galton, branches off towards Düsing and leads on to Fisher. This time the observable variability of the sex ratio at birth is the very fact that must be taken into account. So it is the physiology, the mechanism of reproduction, or the system that characterizes reproduction that must be grasped. As we know, the *statistics of variability* that resulted from this, from Galton's work onwards, brought something very new to the toolkit of those carrying out the calculations. From the mid-20th century, it became the most generally accepted standard for statistical training in biology, economics, sociology and demography.

Let us consider, by way of example, the proportions of boys at birth, S_i, for several consecutive years. From a strictly *probabilistic* point of view, each value means nothing unless it is accompanied by an indication of its own variation: in this case, a confidence interval considered for a certain probability level. The value of S_i can therefore never be taken as certain. In contrast, from a *statistical* point of view, as applied in the calculation, each annual value is viewed as established each year, leaving aside any possible errors, approximations or uncertainties. The use of a statistical method then consists in relating one part of the variations from one year to another to the number for the year, or else to another index that has annual variations, and then in accepting that the remainder of the variability looks "purely" random.

There are two completely different schemas here: firstly, a hypothesis of intrinsic uncertainty from which conclusions can be drawn; secondly, an empirical variability that is decomposed through calculation while keeping the thing we are endeavouring to grasp to something like chance. In the schema that we are describing as *probabilistic*, a chance that is a constituent element of the phenomenon is accepted as such. In the schema that we are marking with the word *statistical*, it is a residual chance, even a discredited one, that measures the variability remaining unexplained once an analytical technique has been applied. It is clear that although these two schemas differ, they are very close in several ways. In the first place, the historical pathways of their formation have crossed and recrossed for three centuries. What is more, up to a certain point, each of them can be conceptualized through the other. A theoretician of the probable will be able to explain important areas of statistical techniques – but not all of them, and not uniformly. A statistician may take the view that certain pillars of probabilistic thinking – for example, the binomial schema that is at work in the analysis of sex ratio – can be sufficiently explained in terms of frequencies.

In the case of the calculation of human sex ratio at birth, a review of three centuries of tensions between these two conceptual schemas also suggests that the relationships between them, as viewed by scholars and applied in their work, have been through several regimes of conditions that made them possible. The concept of a *pre-Quetelesian* regime seems to describe conditions where neither reference to a central value nor analogy between the structure of variability of the phenomenon and the structure of the errors that its measure involves is predominant. Reading D'Alembert, Condorcet, Laplace, Fourier or Babbage without considering them as scholars of such a particular era would lay us open to dangerous anachronisms. Restoring this characteristic would give their texts a depth that historians of the sciences have too blindly reserved for Laplace and too rarely accorded to the other four. The *Quetelesian* regime in its strict sense is better known: in these conditions, the activity of scholars is governed by the normalization of observations according to the theory of the average and of errors, formulated by Quetelet and carried forward by the rise of European statistical institutions. This regime became established during the second half of the 19th century: several authors have already observed a long time ago that the 1840s marked a turning-point in this regard (Daston, 1988; Hacking, 1990; Porter, 1986). But by the end of several decades, the rise of this first form of statistical thinking and the rise of specialized forms of production had had the obvious consequence of highlighting the variability of the phenomenon whose recording had thus been normalized. Most statisticians active after Quetelet were content to reason like him, but using materials whose homogeneity they rejected even though it had been accepted by their predecessor: "average" reasoning was therefore no longer directed at the average man but, for example in the physical anthropology of Paul Broca (1824–1880) and Louis-Adolphe Bertillon, at presumed racial differences (Brian, 1991). So these were instances of Quetelesian thinking in an era which was already no longer exactly Quetelesian. As we have seen, it was in England, and starting from a completely different conception of variability – derived from Babbage and from Darwin, and consolidated in Galton's works, but fostered by Quetelesian statistics – that methods of analysing variance were developed to the point where they would completely change statistics in Continental Europe in the 20th century. Thus it is possible to describe the conditions enabling these later forms of thinking as a *para-Quetelesian* regime. From this arose the works of authors as different as Darwin, Galton, Lexis, Durkheim and Gini – to mention only those who, in one way or another, have appeared in the course of our survey.

In the *pre-Quetelesian* regime, the distinction between probabilistic and statistical schemas does not make much sense. In the *Quetelesian* regime, these schemas are fused into a single doctrine. In the *para-Quetelesian* regime, the texts of the most attentive scholars demonstrate an epistemological tension between the two schemas. The characteristic features of this tension, as we have seen, depend on the specific conditions of the scientific division of labour. Perhaps a schema of thinking that would assume both the probabilistic variability of a phenomenon and its statistical variability without confusing them even by approximation – that is, viewing them as two epistemic dimensions that are *a priori* distinct, even though it combines them analytically at a second stage – should be described as *post-Quetelesian*.[3] This clarification would at the very least have the advantage of providing, on the one hand, a reasoned extension of earlier schemas and, on the other hand (as in the first five chapters of this book), a framework for the critical comparison of past examples. Beyond this critical use, our reconstruction of human sex ratio at birth in Chapter 6 is an example of the empirical scope of this proposition, whose principle lies not in reducing a detested chance but in *adopting* a stochastic hypothesis developed through the calculus of probabilities – in "setting chance against chance" (Condorcet, 1793–1794 [2004, p. 437]; translation from Philadelphia edition, 1796).

[3] In early-20th-century Continental Europe, Corrado Gini and Maurice Halbwachs were among the very rare scholars we are tempted to describe as post-Quetelesians.

Appendix A
CONDORCET (1743–1794)

Extracts from *Effets sur l'état moral et politique de l'espèce humaine de quelques découvertes physiques*[1] (1793–1794)

Bibliothèque nationale, Paris, ms, n.a.f. 4586 folio 189 recto-209 verso; published in Condorcet 2004, pp. 923–936.

Effects on the moral and political state of humankind of some physical discoveries like the means of producing male or female children, as one chooses, with a certain probability; of producing children without the union of the mother with any man; etc., which may have results acting for or against the continued perfecting of humankind.

[1] This *Fragment*, numbered *10* by Condorcet himself, is part of the manuscript that he prepared during his period in hiding from the Revolutionary Terror, with a view to a work entitled "*Tableau historique des progrès de l'esprit humain [A history of the progress of the human mind]*" (all known documents that relate to this have been published in Condorcet, 2004). The posthumously published prospectus, later entitled *Esquisse d'un tableau historique des progrès de l'esprit humain [Sketch for a history of the progress of the human mind]* (Year III – 1795), was intended to herald the *Tableau* – a vast project that was never completed. Léon Cahen was responsible for the first edition of *Fragment 10*, with slight modifications from the manuscript form, in 1922. The publication remained unnoticed for a long time, although in recent decades commentators have most often made use of mere snippets of its complex reasoning in order to extrapolate anachronistic conclusions from them. Therefore readers are warned against the kind of hasty borrowings to which this translation of some longer extracts may well lead.
Condorcet is here in the Tenth Epoch of the *Tableau*, where knowledge of the history of the human mind gives him hope for humanity's future. In his view, therefore, we find ourselves after the advent of equality of rights, for which the American Revolution (1776) and the French Revolutions (1789 and 1792) seem to have definitively opened a new epoch. He attempts to evaluate the advantages and risks of controlling the number of births; the possible effects of a probable action to control the sex of the children to be born; and, finally, the effects of discovering a method for reproducing without sexual union. The text echoes debates of his time, notably those which accompanied the reception of Lazzaro Spallanzani's experiments (see in Chapter 2, on p. 38). Condorcet

In allowing oneself to hope that one might see humankind achieve moral perfection following the continuous, indefinite progress of the Enlightenment, one sometimes feels hampered by the idea that certain discoveries, of which there is nothing to prove the impossibility, may disturb its course towards this achievement of perfection and produce a revolution in the reciprocal relations of human beings that will force them to seek their happiness in contriving new combinations. I will not pay too much attention to the evils that the discovery of new means of destruction would produce. Woe betide anyone who does not feel that, even in times of corrupt and ferocious morals, humankind still lacks these means much less than the will to use them. I shall not dwell on the futile fear of the dangers that would result from the art of travelling through the air,[2] for it can be no secret that this could not multiply the faculty to do harm without increasing the faculty to defend oneself, that it could not help a guilty man to flee without providing the means to pursue him more easily.

But if the perfection of hygiene results in a longer lifespan for human beings, greater fertility and the preservation of a larger number of children; if the perfection of medicine postpones the death of almost all individuals into the oldest ages of life; if this population growth exceeds the limit that the annual reproduction of objects for consumption may reach, will the

---

touches on questions debated at length by economists in the Age of Enlightenment, about the ratios of populations to the means of subsistence. The overall thinking that he presents is fostered by a knowledge of practices for limiting births: that such ancient practices were known to a scholar and a gentleman at the end of the 18th century is today attested by demographic historians and historians of demographic knowledge. Condorcet speculates – not arbitrarily, as if to create a utopia, but as a scholarly conjecture – about a world freed from any form of superstition: a world from which equality of human rights would proceed. He dismisses in one paragraph the prospects of powers of destruction and of what we nowadays call crimes against humanity. Given that superstition would be discarded and human society would be conducted through reason thus freed, Condorcet could only conceive of a new authority being able to take charge of population control techniques. For him, there was no room for a state that would embody, if not corrupt, Reason – quite the opposite of the 19th-century, and then the 20th-century, idea. If this happy vision puzzles us, it is because we are applying an anachronistic reading. For Condorcet, the question of the moral impacts of the ability of humankind to affect the probability of the sexes at birth is resolved in the act of procreation between human beings gifted with reason, whether or not it is assisted by technical methods. Thus, seen in retrospect, his thinking may appear liberal.

[2] This echoes the dazzling reception given in France to the invention of aerostats, the Montgolfier hot-air balloon and the Charles hydrogen balloon in 1783.

human race not find it impossible to escape a destruction that is fatal to its happiness, incompatible with the preservation of the order of societies?[3]

It could also be asked what might be [the effect] on humankind [of][4] the discovery of a means of producing a male or female child according to the will of the parents, a means that must in the end be discovered through careful observations. Supposing that this is likely to become a common practice, that it is sufficiently perfected to give the well-founded hope of success, certain to succeed just once in ten attempts for example, would it [not] lead to [changes] in the social relations of human beings, whose consequences could be harmful to the peaceable development of that indefinite perfectibility with expectations of which we have flattered humankind? And would not these relations undergo a much greater revolution if the birth of a child no longer necessarily supposed the union of its mother with a man?

Rarely have philosophers directed their assured gaze upon those objects located between disgust and ridicule, where both hypocrisy and scandal must be avoided. Christian superstition, preoccupied with the chimeras of a mystical purity, has led us into the habit of attaching ideas to these physical actions, which for good or ill may be applied only to their moral consequences. It would show them only from the point of view of shame or of crime, and one had either to make fun of its risible anger or share it. The custom of treating these questions in a foreign language every time it

[3] This passage is somewhat astonishing, as are several others that develop it in the complete *Fragment* (which, it should be remembered, remained in manuscript form until 1922). Condorcet here tackles, explicitly and in great detail, the objection that occurred to Thomas Malthus after reading the *Esquisse* – where, however, there were only rare allusions to *Fragment 10*. Malthus' criticism of the indefinite perfectibility of humankind related precisely to limits on the means of subsistence: population grows by multiplication and "objects for consumption" by addition, so Condorcet's projection would be faulty. Malthus' *An Essay on the Principle of Population* (1798, 1803) caused a considerable stir. For several commentators – most certainly not very attentive to 18th-century debates – it marked the beginnings of modern demographic thinking. As we can see, Condorcet had envisaged this objection and answered it here. There is no doubt that the geometer would have recognized the mark of superstition in Malthus' reaction and would have attributed to a new kind of priesthood the interventions of those who, since the 19th century, have intended to prescribe people's reproductive behaviours. *Fragment 10* is in effect a manifesto that its author would willingly have described as "anti-superstitious". In this respect, it remains somewhat provocative.

[4] Words missing from the manuscript that were re-established for the 2004 edition are shown in square brackets.

was necessary to talk about them seriously,[5] reserving the common tongue for sermons and jests, leaves almost our only choice between expressions either too scientific or vague and obscure, or given over to the common gaiety. |But although Christian superstition has led us, through exaggeration in the opposite direction, to look at objects that are important to the morals and to the happiness of humankind [only] from the point of view of [?][6] or of jokes, it is time to see them from the perspective presented to us by reason,|[7] time to rise above the hypocrisy of customs and the hypocrisy of style, showing objects in their true light and under their own colours. |All the chains of minds, like those of tongues, must finally be broken.| Why moreover would the philosophers who have braved the league of tyrants and priests fear the league of unpleasant jokers and moral hypocrites? Is their courage to find its limit in fear of ridicule and in the anathemas of a false delicacy or an unnatural austerity?[...][8]

Are the very grounds upon which the moralists have based their prejudices as solid as they seem at first glance? Even supposing that there could be a true obligation not to place any obstacle in the way of the possible birth of a being that one believes must be unhappy, is it an increase in the number of births that is the object of this obligation, rather than a growth in the number of human beings who can accomplish its intention and fulfil its duties? Could there be any other aim than that of multiplying well-formed

[5] The issue of languages is one of the main keys to Condorcet's historiographical vision in the *Tableau historique* and in his *Esquisse*. Condorcet suspected all "priests" – officiants of the cults of antiquity, Christian clerics of all periods, scholars and doctors attached to protecting their knowledge – of hindering the achievement of reason. Their favoured instrument was recourse to the use of an esoteric language. It was in response to this that Condorcet, in another *Fragment* relating to the Tenth Epoch – numbered *4* – was to give a method for the formation of a universal language fostered by the philosophical and scientific experience of the preceding centuries (2004, pp. 947–1029).

[6] Illegible word, possibly "jest".

[7] The passages that appear between two vertical bars are earlier variants of the manuscript, which Condorcet himself crossed out, but which we have re-established in order to clarify his meaning.

[8] A passage follows that develops this, dealing with the capacity of humankind to distinguish the necessities of reproduction from the pleasures associated with sexual acts. We should clarify that birth control practices and certain forms of contraception were known in the 18th century. Condorcet indicates the reasons that seem to him to explain the human capacity to operate this distinction, and the causes of superstitions attached to this phenomenon. In the course of his analysis, he mentions birth control, infanticide by exposure of newborn infants in Rome and in China, contraceptive methods known in the 18th century, adultery, prostitution and "those bizarre tastes, those debasing habits that enervate and degrade humankind" (ms folio 194 verso; 2004, p. 928).

beings capable of being useful to others and of making their own happiness? Will the distribution of children in various families, or the distribution of the times of their birth in each family in particular, better fulfil this aim if it is left to chance than if it were directed by will?

Will it be said that the interest of pleasure is necessary in order for humankind to determine to perpetuate itself? But if children are not the object of their fathers' sweetest hope, if their fathers regard them as an inconvenient burden or as beings condemned to unhappiness by receiving life, this would mean that the generations of humankind are destined to appear on the earth only in order to suffer there. And why then would one occupy oneself with the preservation of a race necessarily miserable and foolish? But if the earth offers all human beings the easy means of providing for their needs, if the social pact ensures them such means, if just laws defend them against wickedness and against oppression, why would the desire to have children not be a universal sentiment, [if] it is dictated by nature and approved by reason? [...]

We may boldly conclude that the epoch where the ways in which humankind made progress – the epoch of the arts, in short that of a greater wisdom in the distribution of cultivated land and of labour – would render all population growth contrary to the general interest, and where at the same time the lifespan of human beings and the preservation of children would bring about this growth if they abandoned themselves to the impulse of nature, that this epoch would not be, for human beings free of all prejudices, a fatal limit where their coming to perfection would have to halt, where humankind would necessarily have to pass through a state of suffering in order to recommence its course towards a new prosperity, only to come to a halt again before the same obstacle. [...][9]

[9] This is a second key to the *Tableau historique*, which derives from the philosophical debates of the Age of Enlightenment. The work is in fact entirely directed at countering *machiavellianism*. This word, as used by Condorcet, does not refer primarily to political calculation, but to the idea that nothing better is to be expected than the current order of the world, and that consequently arbitrary political decisions would be no less just than this order. This machiavellianism is therefore comparable to contemporary political cynicism. In the 18th century, it had to be reconciled with Leibnizian optimism, whose political destiny was to be a form of conservatism (despite the intellectual commotion caused by the Lisbon earthquake of 1755). It was also deeply rooted in a theological view based on the necessary suffering of human beings. For Condorcet, as for Turgot, this system of necessities had been overturned, giving way to a *meliorist* perspective. In his eyes, the belief that human beings are born to suffer results from superstition. In freeing itself from this through the exercise of reason, humanity gives itself the means

In the hypothetical case that one had a means of determining the sex of children at will, at least up to a certain point, would there arise a significant difference in the number of individuals of each sex instead of the almost total equality that exists today,[10] and which then would be the most numerous, and what might be the effects of this disproportion on the social order?

First of all let us examine what grounds could determine a preference for one sex to the other. Up to now, prejudices, much more than reason, have governed the division of labour between the two sexes.[11] This distribution has not even been governed by their almost complete equality of numbers; often there are not enough men to labour, while women remain idle. But we have already examined what the influence of destruction would be on the state of the two sexes. We have seen that especially supposing progress in the art of applying machines to skilled trades, infrequent wars, very long voyages becoming shorter – the necessary consequence of a greater equality of civilization and industry between different peoples – this distribution of labour, left to the will of individuals, would be done in the manner most advantageous to the two sexes according to the ratio of their numbers.[12]

But if a change to this ratio could be mastered, it could then be made with a view to obtaining a distribution of labour more advantageous to the total mass.[13] A distribution exists that is most suitable for each ratio, but between these distributions, the most suitable varying with the ratios, there may be some that are best in themselves.[14] Let us now suppose that the ratio that exists at a given time is less advantageous in regard to the resulting distribution of labour than a certain other subsequent ratio; that in order to move closer to the latter it will be necessary to increase the number of

---

of discerning the conditions for its happiness, first among which ranks equality of rights. Respecting rights ensures the movement of humankind towards prosperity.

[10] Condorcet knew Graunt's and Süssmilch's conclusions, as well as the detail of Laplace's work. Each of them, in his own way, demonstrated that the proportions of the two sexes at birth were close, but from a general point of view always slightly in favour of boys.

[11] This is an analysis in terms of division of labour, emanating directly from the economic texts of Turgot and of Adam Smith, which Condorcet knew extremely well.

[12] Here Condorcet is referring to a passage that has not been retained for this translation and which deals with the population in general, without regard to the distribution of the two sexes. In his view, a "constant parity" reasoning would follow the same routes as the more general precedent. He therefore contented himself with an outline of it.

[13] That is, the whole of humankind.

[14] Condorcet's reasoning on the relationship of cause and effect between the division of labour according to sex and the proportions of the two sexes is completely abstract.

women or that of men; it will be more advantageous to give birth to a girl or a boy. Therefore there will always be a subsisting cause that will act to lead to this more advantageous ratio.

In a society where equality of rights would entail the real equality that is the consequence of it, women would nurse their own children. Thus twelve children may be regarded as a limit that they would surpass only very rarely. But the number of those to which a man may give birth is far from being contained within such narrow boundaries. The result of this is that, where the number of individuals is not enough to labour, the quantity of food easily produced would not be sufficient, so one would seek to increase the number of women, even if this were to result in the establishment of [polygamy].

But this need to increase the population may [not] be encountered in a country that has reached the degree of civilization that we are supposing here,[15] except in extraordinary cases. But if a lesser rate of reproduction is needed, this does not result in a need to increase the ratio of the number of men to that of women. We have seen above that other means exist,[16] and it is difficult to believe that the means of determining the sex at will may be simpler and easier to employ.

If the parents consult the interest of their own happiness or that of their children, this interest will bring them to equality since we are supposing here that reason has led to the disappearance of the institutions that have made independence, freedom, the enjoyment of the rights of humankind, the privilege of one sex alone.

The question therefore amounts to whether the interest of obtaining a more useful employment of the forces of humankind, acting unceasingly and towards a single aim in the midst of other interests that may be opposed to it, must in the end entail a disproportion in the number of individuals. I do not believe so. As women consume less they may, in all employments that may be fulfilled by the two sexes, produce an equal result for the price of an equal expenditure, and in the total mass of labour, of occupations, the portion that can be regarded as the same whether done by either sex is too considerable for there to be any advantage in a disproportion of the

[15] We really are in the "*Tenth Epoch*" here (see above).

[16] This refers to the most varied means then known for limiting births, without regard to their sex.

number, while the portion of labour that it is good to reserve to each sex would suffice to prevent this disproportion from extending beyond certain limits.

But here an objection presents itself. [...] Does not each sex have an interest in multiplying the sex that it is not? Men so that, at an advanced age, liaisons with younger women are easier for them, women so that men, limited to fewer objects, are obliged to become less particular about youth. Here I will observe only that, if that which relates to vanity is set aside, the strength of a long attachment, the propriety of ages, the natural repugnance at being only an object of disgust would destroy this same interest of pleasure in beings directed in general by their reason. I shall not stop to consider that the execution of such an enterprise entails the destruction of that equality of rights which I suppose to have been established; because I would be told that the influence of this very interest would be an obstacle to the preservation of that equality and to the progress of reason. If I add also that the sex which would be reduced to a less [great] number in order to multiply its enjoyment would necessarily become the weaker, I could be told besides that there results from this observation an additional interest in seeking to introduce and strengthen prejudices, to employ those means of oppressing the greatest number, which up to now have had across the whole globe such an awful and lasting success. But I shall observe only that there does not exist here any direct interest for those who may introduce this disproportion, that it would not be advantageous to them within their own family, that success would depend also on the will of that of the two sexes which would lose by this disproportion, and finally that this first success, even supposing that one of the two sexes would ignore its interests or lose its rights, is incompatible with the state in which society is supposed to be.

The only result of this discovery therefore would be a means of re-establishing an almost total equilibrium either in the small portions where nature alone would not have established it or in the extraordinary circumstances where she would have deviated from it.[17]

There would be a real danger if one sex had a pressing interest, opposed to that of the other sex, in augmenting the proportion of the number for itself, but that interest does not exist, and consequently force could not

[17] Thus for Condorcet, where one sex ensured numerical growth relative to the other, the ensuing moral effects would lead to a phenomenon of oscillation in the frequency of births according to sex, which would tend towards the re-establishment of near-equilibrium.

preserve a disproportion contrary to the general progress of humankind once chance had established it[18]. [...][19]

Supposing the individuals of humankind guided by their reason, and by those sweet affections to which nature attaches their happiness, they would not overstep the bounds of an ability from which the interest of those affections and the need to feel their sweetness must distance them. Supposing a social system[20] founded on entire equality of rights, it would be impossible to misuse this ability by force for the unhappiness of a portion of humankind. This ability would therefore be nothing more than a resource employed only in certain particular circumstances| where it would serve to make amends for injustices|. This ability would serve above all to make more intimate and more dear, by finally freeing them from the yoke of necessity, the same relations to which, at first glance, one would have been tempted to believe it posed a danger.[21] Everything that may contribute to making individuals more independent[22] is a good, even relatively to the happiness that they can give each other reciprocally: the more voluntary it is, all the greater it will become.

[18] Condorcet considers that one sex cannot privilege its own reproduction over the long term. An observed deviation would lead to a restoration of near-equilibrium. His reasoning here is combinatorial, envisaging different predilections according to the sex of the agents and that of the children arising from their acts. In all cases, the dynamics of the sex ratio is to re-establish an equilibrium close to equidistribution.

[19] This passage deals with the possibility of procreation without copulation.

[20] The expression "social system" here does not have the meaning that it would today. In Condorcet's writings, as in those of some of his contemporaries, it means the system of rules accepted by people in association. We would now describe this, presupposing a national unit, as a "constitution" – in the legal sense of the word – along with the texts that derive from it. Condorcet wrote " social order" (in the text above, just after the position of our Note 10). In this instance, the meaning is much closer to that of Durkheimian sociology. The expression has connotations of "divine order", since society was then serving as an ordering principle, in lieu and on behalf of the Divinity.

[21] That is, relations between the two sexes.

[22] To be understood as: free from the obstacles of existence and from superstitions.

Appendix B
CHARLES DARWIN (1809–1882)

Human sex ratio, natural selection and sexual selection in The Descent of Man *(1871 & 1874)*

1. The limits of actual enumeration (1871 & 1874)

Hence I was led to investigate, as far as I could, the proportions between the two sexes of as many animals as possible ; but the materials are scanty. [...] Domesticated animals alone afford the |means|[1] of ascertaining the proportional numbers at birth; but no records have been specially kept for this purpose. By indirect means, however, I have collected a considerable body of |statistics|,[2] from which it appears that with most of our domestic animals the sexes are nearly equal at birth. [...] It is, however, in some degree doubtful whether it is safe to infer that the |proportion would be the same|[3] under natural conditions as under domestication; for slight and unknown differences in the conditions affect[4] the proportion of the sexes. (1871, Vol. 1, pp. 263–264; 1874, p. 215).

As no one, as far as I can discover, has paid attention to the relative numbers of the two sexes throughout the animal kingdom, I will here give such materials as I have been able to collect, although they are extremely imperfect. They consist in only a few instances of actual enumeration, and the numbers are not very large. As the proportions are known with certainty |only in mankind|,[5] I will first give them as a standard of comparison. (1871, Vol. 1, p. 300; 1874, p. 242).

[1] Replaces "opportunity" (1871).
[2] Replaces "statistical data" (1871).
[3] Replaces "same proportional numbers would hold good" (1871).
[4] "to a certain extent" (1871) cut.
[5] Replaces "on a large scale in the case of man alone" (1871).

2. The empirical variability of human sex ratio at birth (1871 & 1874)

In England during ten years (from 1857 to 1866) | the average number of children born alive yearly was 707,120|,[6] in the proportion of 104.5 males to 100 females. But in 1857 the male births throughout England were as 105.2, and in 1865 as 104.0 to 100. Looking to separate districts, in Buckinghamshire (where |about|[7] 5000 children are annually born) the *mean* proportion of male to female births, during the whole period of the above ten years, was as 102.8 to 100; whilst in N[orthern] Wales (where the average annual births are 12,873) it was as high as 106.2 to 100. Taking a still smaller district, viz., Rutlandshire (where the annual births average only 739), in 1864 the male births were as 114.6, and in 1862 as |only|[8] 97.0 to 100; but even in this small district the average of the 7385 births during the whole ten years, was as 104.5 to 100: that is in the same ratio as throughout England.[9] The proportions are sometimes slightly disturbed by unknown causes; thus Prof. Faye states 'that in some districts of Norway there has been during a decennial period a steady deficiency of boys, whilst in others the opposite condition has existed.' In France[10] during forty-four years the male to the female births have been as 106.2 to 100; but during this period it has occurred five times in one department, and six times in another, that the female births have exceeded the males. In Russia the average proportion is as high as 108.9 |, and in Philadelphia in the United States as 110.5|[11] to 100.[12] |The average for Europe, deduced by Bickes[13] from about seventy million births, is 106 males to 100 females. On the other hand, with white

[6] Replaces "707,120 children on an annual average have been born alive" (1871).
[7] Replaces "on an average" (1871).
[8] "only" not in 1871.
[9] Darwin's footnote, since 1871: "Twenty-ninth Annual Report of the Registrar-General for 1866. In this report (p. xii) a special decennial table is given". See Registrar-General (1869).
[10] See Bureau des longitudes (1867) and Footnote 12.
[11] "and in. . .as 110.5" not in 1871.
[12] Darwin's footnote, since 1871: "For Norway and Russia, see abstract of Prof. Faye's researches, in 'British and Foreign Medico-Chirurg. Review', April, 1867, pp. 343, 345. For France, the 'Annuaire pour l'An 1867', p. 213." Darwin's addition in 1874 : "For Philadelphia, Dr. Stockton Hough, 'Social Science Assoc.', 1874. For the Cape of Good Hope, Quetelet as quoted by Dr. H. H. Zouteveen, in the Dutch Translation of this work (Vol. i. p. 417), where much information is given on the proportion of the sexes."
[13] The table of Bickes' estimates had been published in the *Zeitung für das Gesamte Medizinalwesen* (February 1831) and in the *Mémorial encyclopédique et progressif des connaissances humaines* [*Progressive encyclopaedic chronicles of human knowledge*] (Paris, 1832–1833). They had been discussed by Villermé (1832*c*) and by Quetelet (1835).

children born at the Cape of Good Hope[14], the proportion of males is so low as to fluctuate during successive years between 90 [47.37%] and 99 [49.75%] males for every 100 females.|[15] It is a singular fact that with Jews the proportion of male births is decidedly larger than with Christians: thus in Prussia the proportion is as 113 [53.05%], in Breslau as 114 [53.27%], and in Livonia as 120 [54.55%] to 100; the Christian births in these countries being the same as usual, for instance, in Livonia as 104 [50.98%] to 100.[16] (1871, Vol. 1, pp. 300–301; 1874, pp. 242–243).[17]

3. The indirect effect of sex ratio at birth on sexual selection (1871 & 1874)

I have remarked that sexual selection would be a simple affair if the males |were considerably more numerous than|[18] the females [...]. For our present purpose we are concerned with the proportions of the sexes, not |only|[19] at birth, but |also|[20] at maturity, and this adds another element of doubt; for it

[14] In 1874, Darwin is referring to additions made by his Dutch translator, Hermanus Hartogh Heys van Zouteveen (1841–1891), to his own work, see Footnote 12.

[15] "The average for Europe...99 males for every 100 females." not in 1871.

[16] Darwin's footnote, since 1871: "In regard to the Jews, see M. Thury, 'La Loi de Production des Sexes', 1863, p. 25". See Thury (1863).

[17] The table below shows the 95% confidence intervals that must be read in association with these figures:

	Number of years	Approximated N	Ratio $M/100F$	Ratio M/N	Confidence int. 95%	Lower benchmark	Upper benchmark
England 1857–1866	10	7 071 200	104.5	51.10%	0.04%	51.06%	51.14%
England 1857	1	707 120	105.2	51.27%	0.12%	51.15%	51.39%
England 1865	1	707 120	104.0	50.98%	0.12%	50.86%	51.10%
Buckinghamshire 1857–1866	10	50 000	102.8	50.69%	0.45%	50.24%	51.14%
Northern Wales 1857–1866	10	128 730	106.2	51.50%	0.28%	51.22%	51.78%
Rutlandshire 1857–1866	10	7 385	104.5	51.10%	1.16%	49.94%	52.26%
Rutlandshire 1862	1	739	97.0	49.24%	3.68%	45.56%	52.92%
Rutlandshire 1864	1	739	114.6	53.40%	3.68%	49.72%	57.08%
France before 1867	44	42 500 000	106.2	51.50%	0.02%	51.49%	51.52%
Europe	1	70 000 000	106.0	51.46%	0.01%	51.44%	51.47%

The local figures are irrelevant, even those accompanied by total numbers of live births. Only the variations, including the annual variations, of ratios for England, France and Europe could show relevant differences.

[18] Replaces "considerably exceeded in number" (1871).

[19] "only" not in 1871.

[20] "also" not in 1871.

is a well-ascertained fact that with man |the number of males dying|[21] before or during birth, and during the first |two|[22] years of infancy |, is considerably larger than that of females|.[23] (1871, Vol. 1, pp. 263–264; 1874, p. 215).

So it will be if the more vigorous males select the more attractive and at the same time healthy and vigorous females; and this will especially hold good if the male defends the female, and aids in providing food for the young. The advantage thus gained by the more vigorous pairs in rearing a larger number of offspring has apparently sufficed to render sexual selection efficient. But |a large numerical preponderance of males over females will|[24] be still more efficient; whether the preponderance |is|[25] only occasional and local, or permanent; whether it |occurs|[26] at birth, or |afterwards|[27] from the greater destruction of the females; or whether it indirectly |follows|[28] from the practice of polygamy. (1871, Vol. 1, p. 271; 1874, pp. 220–221).

4. One possible effect of sexual selection on sex ratio at birth: The excess of male stillbirths (1871 & 1874)

The Male generally more modified than the Female. [...] The great eagerness of the |males|[29] has thus indirectly led to |their much more frequently developing secondary sexual characters than the females|.[30] (1871, Vol. 1, pp. 271 and 274–275; 1874, pp. 221 and 223).

|Prof. Faye remarks that|[31] 'a still greater preponderance of males would be met with, if death struck both sexes in equal proportion in the womb and during birth. But the fact is, that for every 100 still-born females, we have in several countries from 134.6 to 144.9 still-born males. |During the

[21] Replaces "a considerably larger proportion of males than of females die" (1871).
[22] Replaces "few" (1871).
[23] "is considerably larger than that of females" not in 1871.
[24] Replaces "preponderance in number of the males over the females would" (1871).
[25] Replaces "was" (1871).
[26] Replaces "occurred" (1871).
[27] Replaces "subsequently" (1871).
[28] Replaces "followed" (1871).
[29] Replaces "male" (1871).
[30] Replaces "the much more frequent development of secondary sexual characters in the male than in the female" (1871).
[31] Replaces "In various parts of Europe, according to Prof. Faye and other authors," (1871).

first four or five years of life, also, more male children die than females,|[32] for example in England, during the first year, 126 boys die for every 100 girls – a proportion which in France is still more unfavourable.'[33] |Dr. Stockton Hough accounts for these facts in part by the more frequent defective development of males than of females. We have before seen that the male sex is more variable in structure than the female; and variations in important organs would generally be injurious. But the size of the body, and especially of the head, being greater in male than female infants is another cause: for the males are thus more liable to be injured during parturition. Consequently the still-born males are more numerous; and, as a highly competent judge, Dr. Crichton Browne,[34] believes, male infants often suffer in health for some years after birth|.[35] |Owing to|[36] this excess in the death-rate of male children, |both at birth and for some time subsequently, and owing to|[37] the exposure of |grown men|[38] to various dangers, and |to|[39] their tendency to emigrate, the females in all old-settled countries, where statistical records have been kept,[40] are found to preponderate considerably over the males. (1871, Vol. 1, pp. 302; 1874, pp. 243–244).

[32] In 1871, this passage was not inserted in the quotation. Darwin summarized Faye in these terms: "Moreover during the first four or five years of life more male children die than females". Then he returned to the quotation, starting from "for example...".

[33] Darwin's footnote, since 1871: "'British and Foreign Medico-Chirurg. Review', April, 1867, p. 343. Dr. Stark also remarks ('Tenth Annual Reports of Births, Deaths, &c., in Scotland', 1867, p. xxviii) that "These examples may suffice to shew that, at almost every stage of life, the males in Scotland have a greater liability to death and a higher death-rate than the females. The fact, however, of this peculiarity being most strongly developed at that infantile period of life when the dress, food, and general treatment of both sexes are alike, seems to prove that the higher male death-rate is an impressed, natural, and constitutional peculiarity due to sex alone."

[34] Darwin's footnote, not in 1871: " 'West Riding Lunatic Asylum Reports', vol. i, 1871, p. 8. Sir J. Simpson has proved that the head of the male infant exceeds that of the female by 3–8ths of an inch in circumference, and by 1–8th in transverse diameter. Quetelet has shown that woman is born smaller than man; see Dr. Duncan, 'Fecundity, Fertility, and Sterility', 1871, p. 382."

[35] "Dr. Stockton Hough accounts... after birth" not in 1871.

[36] Replaces "As a consequence of" (1871).

[37] Replaces "and of" (1871).

[38] Replaces "men when adult" (1871).

[39] Replaces "of" (1871).

[40] Darwin's footnote in 1871: "With the savage Guaranys of Paraguay, according to the accurate Azara ('Voyages dans l'Amérique mérid.', tom. ii, 1809, pp. 60, 179), the women in the proportion to the men are as 14 to 13." (1871). In 1874, the last words were changed to "the women are to the men in the proportion of 14 to 13".

5. A second possible effect of sexual selection on sex ratio at birth: The excess of illegitimate female births (1871 & 1874[41])

|It seems at first sight a mysterious|[42] fact that in different nations, under different conditions and climates, in Naples, Prussia, Westphalia, |Holland, France, England and the United States|,[43] the excess of male over female births is less when they are illegitimate than when legitimate.[44] |This has been explained by different writers in many different ways, as from the mothers being generally young, from the large proportion of first pregnancies, &c. But we have seen that male infants, from the large size of their heads, suffer more than female infants during parturition; and as the mothers of illegitimate children must be more liable than other women to undergo bad labours, from various causes, such as attempts at concealment by tight lacing, hard work, distress of mind, &c., their male infants would proportionably suffer. And this probably is the most efficient of all the causes of the proportion of males to females born alive being less amongst illegitimate children than amongst the legitimate. With most animals the greater size of the adult male than of the female, is due to the stronger males having conquered the weaker in their struggles for the possession of the females, and no doubt it is owing to this fact that the two sexes of at least some animals differ in size at birth. Thus we have the curious fact that we may attribute the more frequent deaths of male than female infants, especially amongst the illegitimate, at least in part to sexual selection.|[45] (1871, Vol. 1, pp. 301–302; 1874, pp. 244–245).

6. Other possible factors in an unequal distribution of the sexes at birth (1871 & 1874)

It has often been supposed that the relative age of the |two|[46] parents determine [*sic*] the sex of the offspring; and Prof. Leuckart[47] has advanced what he considers sufficient evidence, with respect to man and certain

[41] In the 1871 edition, illegitimate births appear as a quick excursus, following the discussion on the variability of the proportion of the two sexes, not the discussion of stillbirths.

[42] Replaces "It is a still more singular" (1871).

[43] Replaces "France and England" (1871).

[44] Darwin's footnote in 1874: "Babbage, 'Edinburgh Journal of Science', 1829, vol. i., p. 88; also p. 90, on still-born children. On illegitimate children in England, see 'Report of Registrar-General for 1866', p. xv". See Babbage (1829) and Registrar-General (1869).

[45] "This has been explained...at least in part to sexual selection" not in 1871.

[46] "two" not in 1871.

[47] Darwin's footnote, since 1871, with minor typographic alterations: "Leuckart, in Wagner's 'Handworterbuch der Phys.', B[and] iv, 1853, [p]. 774". See

domesticated animals, |that this is one important though not the sole factor|[48] in the result. So again the period of impregnation |relatively to the state of the female|[49] has been thought |by some|[50] to be the efficient cause; but recent observations discountenance this belief.[51] |According to Dr. Stockton Hough,[52] the season of the year, the poverty or wealth of the parents, residence in the country or in cities, the crossing of foreign immigrants, &c., all influence the proportion of the sexes. With|[53] mankind, polygamy has |also|[54] been supposed to lead to the birth of a greater proportion of female infants; but Dr. J. Campbell[55] carefully attended to this subject in the harems of Siam, and concludes that the proportion of male to female births is the same as from monogamous unions. Hardly any animal has been rendered so highly polygamous as |the|[56] English race-horse, and we shall immediately see that his male and female offspring are almost exactly equal in number. (1871, Vol. 1, pp. 302–303; 1874, p. 245).

7. Natural selection and the rebalancing of the sexes at birth (1871)

On the power of natural selection to regulate the proportional numbers of the sexes, and general fertility.[57] In some peculiar cases, an excess in the number of one sex over the other might be a great advantage to a species, as with the sterile females of social insects, or with those animals in which more than one male is requisite to fertilize the female as with certain cirripedes and perhaps certain fishes. An inequality between the sexes in these cases might have been acquired through natural selection, but from their rarity they need not here be further considered. In all ordinary cases an inequality would be no advantage or disadvantage to certain individuals more than to others; and therefore it could hardly have resulted from natural

Leuckart (1853), which highlighted the direction explored by Notter and Hofacker (1827).

48 Replaces "to shew that this is one important factor" (1871).

49 "relatively to the state of the female" not in 1871.

50 "by some" not in 1871.

51 Here, in 1871, Darwin did not reference the argument, but turned his attention to polygamy.

52 Darwin's footnote, not in 1871: "'Social Science Assoc. of Philadelphia', 1874".

53 The sentence "According to Dr. Stockton Hough... all influence the proportion of the sexes" replaces the word "Again", which appeared before the word "with" in 1871.

54 "also" not in 1871.

55 Darwin's footnote, since 1871 : " 'Anthropological Review', April, 1870, p. cviii".

56 Replaces "our" (1871).

57 Section heading, 1871, p. 315.

selection. We must attribute the inequality to the direct action of those unknown conditions, which with mankind lead to the males being born in a somewhat larger excess in certain countries than in others, or which cause the proportion between the sexes to differ slightly in legitimate and illegitimate births.[58]

Let us now take the case of a species[59] producing, from the unknown causes just alluded to, an excess of one sex – we will say of males – these being superfluous and useless, or nearly useless. Could the sexes be equalised through natural selection? We may feel sure, from all characters being variable, that certain pairs would produce a somewhat less excess of males over females than other pairs. The former, supposing the actual number of the offspring to remain constant, would necessarily produce more females, and would therefore be more productive. On the doctrine of chances, a greater number of the offspring of the more productive pairs would survive; and these would inherit a tendency to procreate fewer males and more females. Thus a tendency toward equalisation of the sexes would be brought about.[60] [...].[61]

Nevertheless we may conclude that natural selection will always tend, though sometimes inefficiently[62], to equalise the relative numbers of the two sexes. (1871, Vol. 1, pp. 315–316, p. 318).

8. The puzzle of the *production of the two sexes* (1874)

The proportion of the sexes in relation to natural selection.[63] [...] Besides the several causes previously alluded to, the greater facility of parturition amongst savages, and the less consequent injury to their male infants, would tend to increase the proportion of live-born males to females. [...]

[58] Here Darwin finally mentions directly the major question of the imbalance of the sexes at birth in humankind.

[59] This classic example coincides exactly with the human species.

[60] This second paragraph, which in the end was erased from the second edition, is nowadays read by several commentators as anticipating the conclusions outlined by Düsing (1884) and the principle stated by Fisher (1930).

[61] Darwin goes on to discuss various objections that in the end all come down to the same point.

[62] The case of excess male births in the human species results, in Darwin's view, from this inefficiency.

[63] Section heading, 1874, p. 255.

As the males and females of many animals differ somewhat in habits and are exposed in different degrees to danger, it is probable that in many cases, more of one sex than of the other are habitually destroyed. But as far as I can trace out the complication of causes, an indiscriminate though large destruction of either sex would not tend to modify the sex-producing power of the species. With strictly social animals, such as bees or ants, which produce a vast number of sterile and fertile females in comparison with the males, and to whom this preponderance is of paramount importance, we can see that those communities would flourish best which contained females having a strong inherited tendency to produce more and more females; and in such cases an unequal sex-producing tendency would be ultimately gained through natural selection. With animals living in herds or troops, in which the males come to the front and defend the herd, as with the bisons of North America and certain baboons, it is conceivable that a male-producing tendency might be gained by natural selection; for the individuals of the better defended herds would leave more numerous descendants. In the case of mankind the advantage arising from having a preponderance of men in the tribe is supposed to be one chief cause of the practice of female infanticide.

In no case, as far as we can see, would an inherited tendency to produce both sexes in equal numbers or to produce one sex in excess, be a direct advantage or disadvantage to certain individuals more than to others; for instance, an individual with a tendency to produce more males than females would not succeed better in the battle for life than an individual with an opposite tendency; and therefore a tendency of this kind could not be gained through natural selection. Nevertheless, there are certain animals (for instance, fishes and cirripedes) in which two or more males appear to be necessary for the fertilisation of the female; and the males accordingly largely preponderate, but it is by no means obvious how this male-producing tendency could have been acquired. I formerly thought that when a tendency to produce the two sexes in equal numbers was advantageous to the species, it would follow from natural selection, but I now see that the whole problem is so intricate that it is safer to leave its solution for the future. (1874, p. 255, pp. 258–260).

Appendix C
MAURICE HALBWACHS (1877–1945)

Extracts from *"Recherches statistiques sur la détermination du sexe à la naissance"*[1] (1933)

[1933, p. 164; 2005, p. 381][2]

Perhaps statistics really began when it was recognized that the proportion of births of boys – and of girls – remains more or less constant.[3] But why does it do so? This question has continued to preoccupy statisticians. It is a classic problem, and there is all the more reason to examine it again because, despite more than a century of looking for an answer, the solution can hardly be said to have been found.

A secular trend, with an exceptional variation in the years 1918 to 1920
[1933, pp. 166–168; 2005, pp. 383–384]

The proportion of male births is said to be more or less constant. Even so, let us review it over the whole of the last century,[4] or even for up to 120 years, in France. We shall see that it has fallen slightly: an almost imperceptible decrease from one year to the next, from one decade to the next, but never – at least up until the War[5] – a continuous rise. Since this

[1] We are publishing here extensive extracts from Maurice Halbwachs' article "*Recherches statistiques sur la détermination du sexe à la naissance [Statistical research on the determination of sex at birth]*" (1933), on which we give a commentary on pp. 134–136. The author himself drew on part of it again for a section of the text *Le Point de vue du nombre [The standpoint of number]*, which appeared in the *Encyclopédie française* in 1936. Both versions have been recently republished (2005).

[2] Page references are to the original edition (1933) and the recent new edition (2005), where the reader can consult the detail of these arguments in French.

[3] Halbwachs knew, and mentioned, most of the major works published before his time, from Graunt (1662) to Gini (1908). However, he seems not to have known the manuscript of Condorcet's *Fragment 10*, published by Cahen in 1922.

[4] In 1933, the 19th century.

[5] The First World War.

was a slow decrease, it is possible to calculate the proportion of male births on average over fairly extensive periods, and the deviation, also on average, from that proportion. [...]

Let us now consider the period 1891 to 1929: it is striking that, over this thirty-nine years, the mean deviation[6] of 0.491 was very clearly exceeded in the three years 1918, 1919 and 1920, when the deviations were 1.980, 1.280 and 1.940 respectively. This variation appeared and disappeared abruptly: the two deviations immediately before it were 0.430 and 0.210; the two after it, 0.380 and 0.510. The 1919 deviation is almost three times the mean deviation, while the deviations for 1918 and 1920 are both far larger still, at four times the mean. Not since 1891 – in fact, not even since 1811 – have such large deviations been found; and, in particular, none that have been repeated three years running. [...]

Thus, this is the first time since male and female births have been recorded in France that such a marked variation has been discovered. The only comparable irregularity, which appeared 116 years earlier, was a great deal smaller.

If, in ten or twenty centuries, all memory has been lost and no trace remains of the historical, political and military events that took place in France in the 19th century and the first quarter of the 20th century, and yet a statistical report is found showing, year by year, the ratio of births of boys to births of girls throughout that period of 130 years, it would have to be assumed that in around 1802–1803 – in those years themselves or in the years just before them – and, similarly, in around 1918–1919, some major upheaval must have taken place in the physical or social environment, since the balance in the ratio of births of the two sexes, usually maintained from one year to the next, was so clearly disrupted. It is natural that we should immediately think of the wars of the Revolution and of the Empire and also of the Great War that ended in 1918. But how could such events have altered the ratio of the sexes at birth? [...]

[6] This must be understood as the mean of the deviations, in this case taken on the absolute values of the deviations. We should clarify that Halbwachs was very familiar with the mathematical statistics of his time (see Fréchet & Halbwachs, 1924) and that the use of standard deviations and of tests that are common nowadays became widespread in Europe only from the inter-War period, and especially after the Second World War. This paper, presented to a meeting of the leading French statistical society of that era, is evidence of a knowledge acquired before these became accepted, as well as of a reasonable competence.

A cyclical phenomenon?
[1933, pp. 171–172; 2005, pp. 386–387]

We are [...] led to ask ourselves two questions:

First, if it is true that, from the beginning of the War, the age difference between spouses became smaller, and if in fact that was the cause of the increase in male births, how does it come about that male births increased (in proportion) only from 1918 onwards and not from 1916? We should note that, in normal times, almost half of newborn babies (48%) are first-born children. Let us assume that these first-born children come from marriages that took place in the preceding year. A marked variation in the proportion of boys among first-born children will become apparent in the total. But, during the War, the number of new marriages – which means the number of first-born children – decreased significantly: in 1915 it fell below a third of what it had been in 1913, and in 1916 it remained below half. First-born children therefore represented a much lower proportion of newborns – barely a fifth. Naturally, variations that affected first-born children disappeared within the overall total. In contrast, the number of marriages increased very quickly in 1917, and even more so in 1918 and 1919 (for 100 marriages in 1913, the figures were 65, 73 and 198 respectively). If we add together children born in 1918 of these unions and children born in the same year to marriages that had taken place in 1915 and 1916 (in the same abnormal conditions of parental age gap), their proportion to the total number of newborn babies reaches and even exceeds its pre-War value. This therefore explains why the variation that applies to them (the higher proportion of boys) appears then, and only then;

Second, if it is true that the reduction in the mean age gap between spouses during the War tended to produce a larger relative number of boys, how does it come about that, after 1919 and in the two following years, when this age gap became much smaller still, the proportion of male births fell, returning almost to its pre-War value? In fact, men got married in large numbers immediately after the War. This still left a very large female population, and a significantly reduced male population. Women were forced to marry increasingly younger men. The age gap between spouses was, and was to remain, much smaller than during the War. However, the proportion of boys among newborn babies did not rise, but fell. Thus, after a change in the distribution of households according to age gap, births of boys increase; a bigger change, in the same direction, does not increase them. Must it be concluded from this that the age gap between the parents has no influence on the sex of children?

But it could be that the relationship between the two terms is not a simple one, that the increase of the first up to certain point entails a growth in the second, but that past that point it has an inverse effect. This would be a cyclical movement. This is our hypothesis. Can it be directly confirmed?

Study of birth records in the Bas-Rhin *département*
[1933, pp. 175–178; 2005, pp. 389–391]

For every birth certificate that is issued, register office officials complete an index card showing the ages of the father and the mother (in round years), the sex of the registered child, the number of children born to the same mother both since and before the marriage and their sex. These index cards are sent to the General Office of Statistics for France.[7] We were able to get information from the Office of Statistics for Alsace and Lorraine[8] about the cards completed in 1925 for the two *départements* of Bas-Rhin and Haut-Rhin. We confined our study to the Bas-Rhin index cards. From these, we took the number and sex of live births since marriage (including the child being registered on that occasion) and the age of the parents. [...]

The index cards were grouped into two bundles, corresponding to registrations of male and female births, and then arranged in increasing order of parental age within each bundle. As we could not copy all of them (there were 36,000), we restricted ourselves to following through the whole series in order of presentation, keeping one index card out of four (the first, the fifth, the ninth, etc).[9] Thus there was a good chance, in proportion, that the cases we kept would number about the same in each category as in all the cases. [...] It seems likely [according to our study of these records] that the proportion of male births does not vary with absolute age, but with parental age difference. [...] Given the method that we followed (methodical

[7] *Statistique générale de la France* (S.G.F.), now the French National Office of Statistics, *Institut national de statistique et d'études économiques* (I.N.S.E.E.).

[8] Halbwachs was a professor at the University of Strasbourg, in Alsace. The Office of Statistics referred to, which was specifically for the territories regained by France from Germany after 1918, was then the only regional office of statistics in France. Its infrastructure and expertise derived directly from the German experience in the period 1871–1918.

[9] This technique, known as "methodical randomization", was known to German and French statisticians of the day.

randomization), it seems likely that a review of all 36,000 births would have given results very much in the same neighbourhood.[10]

On the other hand, if we start from Category 0 (both parents the same age) – at which the ratio reaches its maximum – and follow the variation either in the negative direction (mother older than father) or in the positive direction (the opposite), we observe that the variation is the same; in other words, we find almost the same proportion of male births for a given age gap, whether it is the father or the mother who is the older of the two. And the same is true (in the three categories that give us an adequate number of male births) whether the households are young ones or older ones. [...]

One is sometimes struck by the fact that, in a family with four, five or six children, all are the same sex, and one wonders if there are "families who have boys" and "families who have girls". The index cards we studied were, as we have said, arranged in two bundles – registrations of boys and registrations of girls. It could be assumed that, if there are indeed "families who have boys", there is a good chance that a larger number of them will be found in the first bundle. So, for each of the two categories, we calculated the proportion of boys by eliminating the last birth (in the first category, all these were male; in the second, all female). We found that the proportion of male births was very obviously the same in both categories. We re-did the same calculation using the data from our next, much more extensive survey, and got the same result. Therefore, there is no reason to suppose that, if the first child, or the first two, are boys, the others will be the same sex; similarly for girls. [...]

Records of allowances paid to large families in the Bas-Rhin *département*
[1933, pp. 178–183; 2005, pp. 391–395]

How could we find out, alongside the number and sex of the children, not just the age of the parents in round years, but their date of birth, in order to be able to calculate the exact age gap, in months, between the father and the mother?

In France, for some ten years, every time a newborn baby is registered, register offices have been recording the father's and mother's dates of birth. But it would take a long time to peruse all these registers. Even if we gave up the idea of finding all the children born to each household (which would

[10] We should note that Halbwachs does not give the confidence interval here, even though, from a technical point of view, he was in a position to do so (see Fréchet & Halbwachs, 1924).

be difficult, except in small municipalities),[11] we would have to look at all births indiscriminately over a period, copying the parents' dates of birth for each one. This would be the best method: but we would need a whole team of researchers. [...][12]

Then there were the allowances awarded by the *départements* to large families for some years. This time we were lucky enough to find a large series of records for the Bas-Rhin *département*, from which we were able to draw the following data: 1st the number of children of each sex, for each household with at least four children, legitimate or legitimated, still living.[13] The sex of each child is not indicated, but the forename is. We discarded cases where the forename could be that of either a boy or a girl;[14] 2nd the father's and mother's dates of birth. These are copied onto the form completed by the mayor of the municipality, on his own responsibility. "The undersigned mayor certifies the accuracy of the above declarations, having verified the documents presented and in particular the official family record book." The father's and the mother's birth certificates usually accompany the record of the allowance paid, so – as far as possible – we compared the certificates themselves with the form completed by the mayor and discarded all cases where the two registrations did not agree. It is possible, on the basis of the date of the marriage, which is also recorded, to exclude children born before the marriage, illegitimate children (who anyway are not taken into account when calculating the allowance) and children born to the same mother or to the same father, but in a previous marriage.[15]

Besides, children born to the same mother, but not the same father, do not have the same surname. Where children were born to the same father

[11] Halbwachs did check this for four "small municipalities" in different parts of France; he gave the preliminary results of this survey earlier in the article, in a section that we have not included here. The "best method" that he describes would be developed by others: in France, after the Second World War, it would take the form of the family reconstruction technique used in the Henry Survey in historical demography (Rosental, 1996; Séguy, 2001).

[12] In this era, university academics in France – and especially philosophers, among whom were counted sociologists – did not enjoy research facilities like those which developed after the late 1930s.

[13] There is a bias here. Halbwachs attempts to deal with this objection further on.

[14] This is a new bias, but Halbwachs does not return to this one. As a Durkheimian sociologist, with a close interest in systems of representations, he ought to have been concerned about it.

[15] Implicitly, Halbwachs studied only the couples' legitimate offspring. Here again, the object that he wanted to capture was reduced according to a principle of which, as an author, he was not very critical. In fact, throughout the article, Halbwachs uses a series of arbitrary cross-sections, all capturing the same object from different angles.

in an earlier marriage, this was most often mentioned expressly. This type of study, as can be seen, demands some attention, and is not always free of the risk of error. This was an additional reason to extend our investigation to the largest number of cases that could be obtained.[16]

The Archives Office of the Bas-Rhin *département* gave us access to these records of allowances, with the permission of the prefect. But we cannot describe our study as archive research. This is because, in order to include a large number of cases, we had to restrict ourselves to keeping only the numerical data that we wanted to use from each record; and therefore we ignored, among other items, the date of the marriage, the dates of birth of the successive children and their birth order, as well as the parents' names, place of residence and occupations, and all the information about their financial situation.[17] We started from the oldest files and followed the whole series in order of presentation up to the most recent files, without any break, over a period of about five years. There were certainly some households that we found several times, but with a different number of children, so that we were able to treat them as new cases; moreover, they represented only a very small proportion of the whole. In total, taking only the cases we actually used, our survey focused on 50,561 births for age gaps from −30 months (where the woman was older than the man) to +132 months, or from $-2^1/_2$ years to +11 years, out of about 56,500 that we found for all age gaps.

We devoted more than three hundred hours, spread over a year, to this study.[18] Given that each file gave us on average five births, we would have needed about five times as long if we had had to take the same number of pieces of data from entries in registers, where only one birth is recorded on each occasion. The fact that no one has done the same work before is probably because this source no longer exists, and – when it did exist – no one dreamt of using it for this purpose.

We should now anticipate two objections. Firstly, we confined ourselves to large families and to somewhat older households. But it has long been believed

[16] Loyal to Quetelet, Halbwachs has great faith in a law of large numbers and, like so many of his contemporaries, underestimates the risk of systematic errors – although he does have some idea of this.

[17] Here Halbwachs, as a sociologist, is lamenting the considerable reduction in information resulting from the use of a statistical technique. This topic touches on an issue that is very much alive nowadays in discussions of method in the social history of populations.

[18] Like his German counterpart Max Weber and the specialists of his day, Maurice Halbwachs worked "by hand" and himself performed the most humble tasks required by his science.

that the absolute age of the parents has no influence on the proportion of births of each sex, and the results of our previous survey did seem to indicate this. [...] The second objection is more serious. We know that, although more boys are born, more of them also die in the early years, so that the balance is re-established after some time, and there are even slightly more girls in total.[19] Our households, at the time we found them, had at least four children. We do not know the number or sex of those who had died since the birth of the first. But the higher mortality of boys emerges from the fact that the mean proportion of boys, for our whole total, was 101 (to be exact, 100.85) per 100 girls [50.21%], as opposed to 105.5 [51.34%] on average in Bas-Rhin for the four years 1925–1928. This difference does not pose any danger to large numbers; but where we are working on categories that contain only a small number of cases, that is no longer true. There is only one means of negating this cause of error: to increase the number of observations. To compare the results of our survey with others where births (and not children still living at the present time) have been counted, it is necessary to multiply the proportions that we found by 1.04 (the ratio of 105.5 to 101). If we confine ourselves to the current data, the results of comparisons established between them retain all their value, since all the items are equally affected by the same cause.[20]

[...W]e should ask ourselves whether they confirm or refute the Hofacker-Sadler theory,[21] so much decried nowadays, and whether it is correct, as those authors believed, that when the father is older male births increase, and when the mother is older, the opposite.

Parental age difference (in years)	Boys	Girls	Number of boys per 100 girls	[95% confidence intervals][22]
−2.5 to 0[23]	3 333	3 348	99.55	[48.66%–51.11%]
0 to 11	22 056	21 824	101.05 [101.06][24]	[49.79%–50.74%]

[19] The idea that "the balance is re-established" is presented here rather simply, and in an intellectual tradition going back to Süssmilch (1741). The phenomenon is actually much more complex. See in Chapter 1 p. 6.

[20] This was how Halbwachs hoped to escape the empirical biases that his sources entailed.

[21] See above, pp. 66–69.

[22] This column, which does not appear in the 1933 or 1936 editions, is ours (on this statistical criterion, see in Chapter 5, pp. 134–136). The same is true for the corrections to Halbwachs' rounding, shown in square brackets.

[23] None of the couples studied consisted of parents born on the same day. Here and on the line below, it must be understood that the case "0" is excluded to within a day.

[24] Rounded incorrectly.

It can be seen that this result conforms to the Hofacker-Sadler law: there is a difference between the two proportions of male births (older father, older mother) of 1.50 in favour of households where the father is the older.[25]

These two statisticians further added that the proportion of male births was higher still when both parents were the same age. Let us take parents where one is no more than a year older than the other as being the same age, and let us calculate the proportion of male births for the following age differences:

Parental age difference (in years)	Boys	Girls	Number of boys per 100 girls	[95% confidence intervals][26]
Below −1	1 329	1 433	92.74	[46.21%–50.02%]
From −1 to +1	4 941	4 765	103.70 [103.69]	[49.89%–51.92%]
Below +1	19 119	18 974	100.75 [100.76]	[49.68%–50.70%]

We can, therefore, say that this second part of the Hofacker-Sadler law is also confirmed,[27] and, with these new total numbers, the difference between cases where the father is older and opposite cases appears much more marked than with the preceding numbers. It is true that, for age gaps below −1 year, we have only a limited number of births. But this is probably the first time that it has been possible to subject the law in question to precise, detailed confirmation. Moreover, we shall see that it corresponds to a very general view of our phenomenon, and that the predominance of boys cannot be explained solely – nor even especially, as these two statisticians believed – by the fact that households where the father is older than the mother are the most numerous. [...]

Relationship between the proportion of births of boys and parental age difference
[1933, pp. 184–191; 2005, pp. 395–400]

From now on, let us confine ourselves to the detailed series of numbers (proportion of male births) calculated according to age differences increasing by six months at a time.[28] [...] No doubt the curve that represents such a

[25] 1.50 is the difference between 101.05 and 99.55. But the conclusion is wrong, as the overlapping 95% confidence intervals in the right-hand column of the table show.
[26] Same principles as in the previous table.
[27] This new conclusion is wrong, for the same reasons as the preceding one.
[28] This criterion for defining age gap intervals is a particular feature of Maurice Halbwachs' work. He returns to its importance at the end of the article.

movement includes many irregularities: the maxima are in more or less the same neighbourhood, but one of the three minima is far below the others; on the other hand, [a] first cycle extends across three years, while the two others each extend over only two; and it seems that as we move towards the bigger age differences, the period narrows even more. However, this is a cyclical curve, and, if it is imperfect, that is because we are in the sphere of life whose complexity most likely does not yield to the regularity of mathematical expressions. In any case, it can be stated that the results of a game of chance would not line up in this kind of order.[29] These variations are *real*. There is certainly a relationship between the values taken by the proportion of male births and the age difference between the parents. In the complex problem of how sex at birth is determined, this is the only factor whose effect we have, up to now, been able to recognize and measure.

Let us now recall the hypothesis that we formulated after studying the ratio of male to female births in France throughout the War and up to the most recent years. We noted that the value of this ratio increased abruptly and very markedly in the three years 1918, 1919 and 1920, following the wartime period during which the mean age difference between spouses had become markedly smaller. Was the latter the cause of the former? However, after 1920 and for about ten years thereafter, the mean age difference between spouses became even smaller still. But the proportion of male births – instead of rising – fell, returning almost to the pre-War level. Was there therefore no relationship between parental age gap and this proportion? Or should we instead assume that this relationship was subject to variations such that, two terms being linked, the fall in one to a certain point entails an increase in the other, but once this point is passed, it causes a decrease? We therefore assumed that this relationship perhaps varied in a kind of cyclical function. Direct observation of births, distinguished by sex, according to increasing parental age difference, teaches us that the ratio between the two terms – parental age difference and proportion of male births – does indeed vary as a cyclical function [...]. It seems, therefore, that the facts we have just described, as we were able to observe them, confirm our hypothesis. [...]

Having noted that in 1918, 1919 and 1920 the proportion of male births rose in an apparently abnormal way, reaching values that had not been encountered during the preceding decades, and that nor, moreover, had

[29] Here again, the conclusion is conceptually admissible but lacks any measured criterion. It can be shown that the fluctuations highlighted by Halbwachs remain within an uncertainty band that corresponds to a 95% confidence interval for each age gap group, and that it is futile to try and save the "Hofacker-Sadler law" (see 2005, pp. 172–175).

Ratio of male births to female births according to parental age difference (from 50,561 births in the Bas-Rhin *département*)

Courtesy of I.N.E.D. (published in Halbwachs, 2005, p. 396)

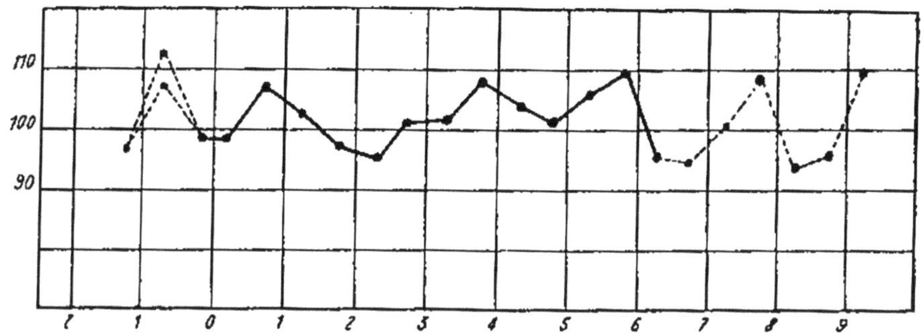

The horizontal divisions represent parental age differences in years – positive ones to the right of zero and negative ones, where the mother is older than the father, to the left. The points on the curve are positioned in the middle of the intervals corresponding to six-month increases in these differences (from 0 to 6 months, from 6 months to 12, etc.). The vertical divisions represent the proportion of births of boys per 100 girls. On the left of 0, the lower dotted line corresponds to the values found for a slightly smaller number of births.

any variation of this size occurred during the whole preceding century, we wondered whether this proportion was in line with a corresponding variation in the mean parental age difference. In fact, during the War, the age gap between spouses became markedly smaller. But it diminished even more after 1920, remaining very low until recent years; however, from 1921 onwards, the proportion of male births has fallen, tending to settle at the pre-War level. We can therefore formulate the following hypothesis: when parental age gap decreases within certain limits, the proportion of male births rises; the proportion of male births falls when this limit is exceeded, and when the age gap becomes far smaller still. A series of direct investigations – the last of which looked at 50,500 births, which we were able to classify by sex according to a parental age difference increasing by six months at a time – enabled us to recognize that the relationship between these two terms could be represented as a cyclical function, with the proportion of male births progressing continuously through a sequence of maxima and minima as the parental age difference increases. Thus, our hypothesis seemed to be confirmed.

More generally, there would be good reason to take into account anew the theory – formulated almost a century ago, and criticized and dismissed by statisticians who have studied this problem most recently – that parental age difference exercises a very clear influence on the proportion of male births. Although it was believed that this was belied by the facts, that was because statistics were used that showed this proportion only for parental age differences increasing by five years at a time. In such averaged figures, the real variations disappear: they reappear when these differences are increased by smaller quantities – of one year and, especially, of six months.

The facts that we observed were most certainly limited. Our final piece of research related to a single *département* and looked at only some of the births that had taken place there. However, it seems to us that the only way forward towards a solution is through a precise, limited study, in which we do our best to look beneath averages that correspond to some very large numbers but are confused, to the realities they conceal. [...]

The ratio of the sexes at birth is not simply a curious fact. It poses a very big problem. We do not under-rate the importance of the calculus of probabilities and of its applications in the sciences. But, if births were comparable to chance facts, then one of the essential functions of organic life would be subject to purely mathematical or mechanical laws.[30] Of course, in a sense, it is a matter of chance whether a boy or a girl is born, and there is perhaps a profound symbolic truth in the old notion that births were related to the conjunction of the stars.[31] But a given age distribution within

[30] This anti-mechanistic passage echoes the vitalism of Henri Bergson, of whom Maurice Halbwachs was an attentive pupil. However, in the first place it arises from the sociologist's adherence to the conception of the social fact by his master, Émile Durkheim (1895).

[31] Laplace, Quetelet and Durkheim are here in the background, and this "profound symbolic truth" results from the analogy between two combinations of chances, general laws and subjective lack of comprehension. Indeed, Durkheim wrote about one configuration of transmission of totems: "here, the totem of the child is not necessarily either that of the mother or that of the father; it is that of a mythical ancestor who came, by processes which the observers recount in different ways, and mysteriously fecundated the mother at the moment of conception. A special process makes it possible to learn which ancestor it was and to which totemic group he belonged. But since it was only chance which determined that this ancestor happened to be near the mother, rather than another, the totem of the child is thus found to depend finally upon fortuitous circumstances" (Durkheim, 1912 [1979, p. 150]; translation by Swain, 1915/1965).

a group is not a matter of chance; and if this distribution explains, at least in part, the ratio of the sexes at birth, then the birth, in a given society, of a given proportion of boys is not a matter of chance either.[32]

[32] Although Halbwachs' result on the "cyclical" dependence between parental age gap and the probability of the children's sex cannot be accepted, this concluding sentence remains pertinent (see above, Chapter 6).

Appendix D

SEX RATIOS AT BIRTH AND THE CALCULUS OF PROBABILITIES

This Appendix aims to clarify a common thread that joins several technical aspects of the study of sex ratios and to provide a common basis for criticizing them according to criteria derived from mathematics, sociology or biology, or from the history of these sciences. We are well aware that a specialist in each of these different fields might take the view that it is "too much or too little". Nevertheless, this note remains necessary in facing our principal difficulty: the fact that collective memories have differed between the various disciplines for three centuries and, in that time, they have already been subjected to various combinations – so we must give careful consideration to any linkages that we want to construct between them.

1. Two traditional indicators

Two indices that express the regularity of ratios between numbers of births according to sex have run through the literature for a long time.

• The ratio of chances of the two sexes

Graunt (1662) and Süssmilch (1741) used the number of boys per hundred girls in a given place and within a given period (for example, 105 boys per hundred girls). Letting M be the number of male births, F the number of female births, N their total ($N = M + F$), this rate is $100M/F$ (called Γ from now on).

Laplace (1778) also employed the ratio M/F (for example 41/39). Conforming to the older usage of considering the chances of different possible cases, Cournot (1843) interpreted this ratio as the number of *chances* of giving birth to a boy to the number of *chances* of giving birth to a girl.

• The proportion of one of the sexes

This time, looking at the proportion of boys among births, $S = M/N$ *or* $M/(M+F)$ (or, for the proportion of girls, taking $1 - S$). This index was sometimes used by Laplace (1778), but most systematically in Poisson's work (1830). They both made use of decimal notations (e.g., 0.5125), whereas today we most often use percentages (in this example, 51.25%). This represents the *frequency* of male births. Cournot (1843) recognized here a *probability* in the sense of a value taken between 0 and 1, which expresses the possibility of the birth of a boy.

2. Historical Markers

Since Jakob Bernoulli (1713), Laplace (1778, 1812) and Poisson (1830), who used mathematical proofs that involved sometimes different conceptions of the mathematics of chances, the *frequency* of these observed events has been linked to the *probability that can in principle be assigned* to each occurrence of these supposedly similar events (the birth of a boy, for example). Several 18th-century authors commented on this link. Thus, Laplace set out to measure the *probability* of a *cause* (that is, the greater or less facility of births of boys) by counting the *events* that it produces (that is, actual births).

Between 1772 and 1830, Laplace (1778), Condorcet (1786), Laplace again (1812), Fourier (1821) and Poisson (1830) worked within an area where measurement error calculation and calculation of the regularity of observed frequencies intersected. Quetelet (e.g. 1846) proposed the view that the two questions were analogous – which simplified their solutions, at least for practical purposes.

The calculation of the proportion of one sex among births was the topic favoured for all these discussions. This scenario is sometimes compared to a series of games of heads or tails (with the coin loaded or not), sometimes to a series of blind throws of two types of balls (or of tickets or of beans) placed in the same urn. The proportion of balls of each type (of each colour, for example) could be balanced (50%–50%), imbalanced or unknown. So many stylized experiments helped scholars to make their proposed calculations understood.

From the mid-18th century, scholars became aware that it was dangerous to make a direct comparison between something concrete that could be counted and a game of chance. The works of Laplace and of Condorcet in the 1770s and 1780s, and those of Gauss, Laplace, Fourier and Poisson during the early decades of the 19th century helped to strengthen an *analytical* conception of the calculus of probabilities, based on integral calculus processes and no longer simply on the relationship between possible cases counted and favourable cases counted. We should add that the mathematics of chances did not end there. In the 20th century, several approaches were taken to renewing them: an axiomatic conception based on the topological theoretic of measure was substituted for certain sections of Laplace's calculus; from the inter-war period, there was a proliferation of mathematical statistics – an approach that cannot always be reduced to the axiomatization of measure; and finally, the more recent development of computerized statistics has sometimes been successful in highlighting regularities whose axiomatic or mathematical proof is still the subject of contemporary research.

• **The *classical* calculus of probabilities**

Throughout these developments, from the mid-18th century to the early decades of the 20th century, counting the sexes at birth offered an exemplary empirical trial. This topic was discussed in three ways, and it is useful to distinguish between them. Firstly, many and varied numerical observations have been available since the 17th century, for reasons that have not always been scientific in nature (parish registers; tax censuses). Secondly, academic thinking about the issue continued to develop without necessarily making use of the calculus of probabilities (medicine, moral sciences). Lastly, the schema of two mutually exclusive possibilities lent itself to *binomial* mathematical calculations.

Since the 17th century, it has been known that the numbers of combinations in such a case may be obtained by developing a binomial of the type $(a+b)^n$. Supposing that there are only two possible cases (a given sex being identified – whether boy or girl does not matter – and its alternative imagined respectively as girl or boy, on the understanding that for this reasoning, it is enough to view one sex as identified on one hand and all other cases on the other hand), and that the first has a probability s and the second a probability s', we have by construction $s+s'=1$ (1 being taken for the measure of certainty). For N cases, this certainty remains 1 and is written $1=1^N=(s+s')^N$, so that:

$$1=(s+s')^N=s^N+\cdots+\binom{N}{n}s^n s'^{(N-n)}+\cdots+s'^N$$

Here $\binom{N}{n}$ is the number of ways of taking n elements (without worrying about their order) in a set that includes N. The preceding formula amounts to the same thing as saying that this binomial can be broken down according to all possible combinations of the case in question and of its alternative. These combinations number 2^N.

The development of the binomial expresses the fact that *insofar as this is a matter of counting, anything can happen*: n cases of “boys” (from now on we shall use this category in the conventional way) and $(N - n)$ cases that are not identified as “boys”, for all the values of n taken from 1 to N. From a strictly combinatorial point of view, case n will occur $\binom{N}{n}$ times and its probability is therefore $\binom{N}{n}/2^N$. Here we may speak of the *classical calculus of probabilities*, characterized by this ratio of $\binom{N}{n}$ favourable cases to 2^N possible cases. This was the reasoning used in the 17th and 18th centuries.

- **The *analytical* calculus of probabilities**

Various pieces of work in the final decades of the 18th century (Hald, 1990; Todhunter, 1865) prepared the ground for this, and it was introduced by the young Laplace; it is characterized by a fairly *similar* construction of probability, but this time on the level of the differential element that then has to be integrated according to the calculus operation. Laplace and his contemporaries devoted themselves to it at the cost of laborious development and new reflections on the theory of functions (hence the adjective *analytical*). From the time of the tensions between Laplace and Condorcet, this analogy has been the object of discussions, still bringing mathematics and philosophy together today (Brian, 1994a).

At this point, a historical sociological issue is not without importance. Although Laplace’s calculus, its developments and its revisions led mathematicians out of some theological ruts, mastery of all these called for a particular mathematical competence, unfortunately most often routinized to established calculation techniques. The result was that its fundamentals were not so familiar to 19th- and 20th-century statisticians and philosophers, even though some of them had come into contact, in one way or another, with the bases of differential and integral calculus. Authors who knew how to control both the calculus and its philosophical analysis were rare – a fact that made the works of Condorcet (1783–1787), Laplace (1814), Cournot (1843), Venn (1888), Bertrand (1889), Borel (1924) and Fréchet and Halbwachs (1924) all the more valuable.

Let us consider that one case (let us say, the birth of a “boy”, for example) has a probability p. Let q be the probability of the alternative to this case: $q = (1 - p)$. Next, possible events must be envisaged, always according to the principle that *insofar as this is a matter of counting, anything can happen*, which is to say that all the proportions observed *a posteriori* x

and $(1-x)$ are possible, with x varying from 0 to 1 (Todhunter, 1865, e.g., §896–897, §1025, §1031).

The following expression then gives a basis for measurement. Once it is understood that the probability of an arrangement of events independent of one another is the product of the probabilities of its components, the formula conveys that, from x to $x + dx$, the share of probability to be taken into account is the product of x in proportion to p and of $(1-x)$ in proportion to q.

$$\int_0^1 x^p(1-x)^q \, dx$$

The calculus of probabilities, whether classical or analytical, characteristically starts from the fact that even though a thing cannot be taken as certain, one can still think very rigorously about its uncertainty. The calculus is then based on taking into account the whole spectrum of possibles, weighted by a measurement of each possibility (Condorcet, 1783; on this topic, Brian, 1994a). In both cases, the measurement of an event is obtained by comparing it with the arithmetical sum of all the possibles (classical calculus) or with the integral on all the possibles (analytical calculus), which is not exactly the same thing. The fact that integral calculus was decisive from Laplace onwards was because, compared to simple addition, it allowed processes that were more subtle, less ambiguous and convenient to approximations (which were almost always indispensable).

3. Sex ratio and probability: notations

In order to clarify this thinking, nowadays it is necessary to distinguish between several different things and then attribute different notations to them. These notations are used below, and again in the development of Chapter 6.

- **Notation** N. – The total number of observed cases, taking both sexes together. M will be the number of boys; F, that of girls; i, the index of any case (taken from 1 to N).
- **Notation** s. – The probability of one sex or the other at a particular moment of ontogenesis. For a case under consideration at the point of conception or even at birth, this is – for example – the *abstract* probability that the case is one of a boy. Laplace, in his texts on the proportion of the sexes at birth, saw this as a physical cause which he called the "*greater facility*" of male births.

- **Notation γ.** – The binomial *random variable* without which it would not nowadays be possible to reason mathematically on the possibility that a child is one sex or another. This variable conventionally takes the value 1 if it is a matter of one sex, 0 if the alternative. In the following, we shall associate the value 1 with the male sex. For a birth i, we let this random variable be γ_i. The parameter s is characteristic of this binomial if s is the probability that it takes the value 1. We may write this as: $\text{Proba}(\gamma_i = 1) = s$. Then the mathematical expectation of this random variable γ_i is also equal to s. This is written as: $E(\gamma_i) = s$.
- **Notation S.** – The *observed secondary sex ratio* – that is, the observed frequency of the male sex among live births. $S = M/N$.
- **Notation Σ.** – The *random variable* that empirical measurement of this frequency S achieves for a set of N cases. It is made up of N elementary random variables of type γ thus: $\Sigma = (\gamma_1 + \cdots + \gamma_N)/N$. Its value, always random, lies between 0 and 1.
- **Notation Γ.** – The *number of boys per hundred girls*, the indicator used since the 17th century. $\Gamma = 100\,M/F$.
- **Notation Π.** – The probability that S is within a certain interval between two given values. The frequency S is, from the point of view of the calculus of probabilities, the realization of a random variable of type Σ. Therefore the observed frequency, the sex ratio S, and any value s can be compared by asking what the probability is that S and that value are sufficiently distinct from each other. This is the question of *estimating* Σ using S, with Σ understood as random and S as observed. Since Jakob Bernoulli (1713), the notions attached to the definitions of s, S, N and Π have been linked by a class of proofs identified (from the 19th century) by the expressions *law of large numbers* or *central limit theorem*. These allow a *confidence interval* to be drawn around S, characterized by a certain amplitude at a given level of probability Π. Applying the calculus of probabilities then consists of rigorously establishing a sentence of the kind: "the probability that the absolute difference between S and the mathematical expectation of Σ is greater than a given quantity is less than Π".

4. Probabilistic estimation of the secondary sex ratio

If the hypothesis is that, for the N cases observed, distribution of the random variable γ is constant and uniform – $\forall\, i$, $\text{Proba}\,(\gamma_i = 1) = s$ – and that these laws γ_i are independent of one another, how is the link between the frequency of observations S and the probability s to be measured? Central limit theorem-type results lead to the conclusion that S is a good estimate of s; that is, that the law of probability of the frequency S rapidly approaches – from the point when N exceeds about ten – a law whose formula can be established. This is the "second law of Laplace" (also known as Laplace-Gauss distribution, or *normal distribution*). Its mathematical expectation (the parameter of its centre) is precisely s, and its variance (that of its dispersal around this centre) is σ, such that $\sigma^2 = s(1-s)/N$. This density is written as:

$$\frac{1}{\sigma\sqrt{2\pi}} e^{-\frac{(x-s)^2}{2\sigma^2}}$$

It is not necessary to take into account the fact that the variance itself depends on s. This is because the value of s is always between 0 and 1, and therefore $s(1-s)$ is smaller than 1/4. Therefore:

$$\sigma^2 < \frac{1}{4N} \quad \text{and} \quad \sigma < \frac{1}{2\sqrt{N}}$$

The density formula (which is already an approximation) and this last upper bound allow two series of values α and Π, which obey this relationship, to be linked:

$$\text{Proba}\left(|s-S| \geq \frac{\alpha}{2\sqrt{N}}\right) \leq \Pi$$

For a long time, tables were used to link α and Π. Nowadays, it is easy to do this using spreadsheet software. It is convenient, although completely arbitrary, to limit ourselves to the pair $\alpha = 2$ and $\Pi = 4.55\%$, and although it is a misuse of language, it is acceptable to take $\alpha = 2$ and $\Pi = 5\%$: we can then speak of a *95% confidence level.*

$$\text{Proba}\left(|s-S| \geq \frac{1}{\sqrt{N}}\right) \leq 5\%$$

This formula is very commonly used. It explains why Fourier (1821) considered it sufficient, from a practical point of view, to confine himself to the fact that uncertainty about the proportion of male births was in the order of $1/\sqrt{N}$. Although a reduction in uncertainty of frequencies according to the number of cases considered had been recorded from the work of Jakob Bernoulli (1713) onwards, it was only following Poisson (1830) that the mathematical concepts involved were clarified.

The preceding formula links the number of observed cases N and the amplitude of the probable difference between the probability s and the observed frequency S. One difficulty in understanding this method arises from the fact that there are *three* things here that all come under the heading "probabilities": first of all, the presupposed probability s, then the frequency S observed *a posteriori*, and finally the probability of observing a difference between the two. This is the price to be paid for thinking rigorously in the face of an intrinsically uncertain phenomenon.

Thus, in order to judge a deviation of 1% on a proportion of boys at birth, with only a 5% chance of being mistaken, at least 10,000 observations are necessary. A deviation of 0.1% would call for a total number in the order of a million cases. Unfortunately, it is clear that many studies published nowadays commonly – and this despite Poisson's very explicit recommendations (1830) – offer estimates of proportions of births that are highly uncertain because numbers of cases are too low. The empirical strategy followed in Chapter 6 of this book is to deal with very large total numbers (often hundreds of millions), in order to reduce the random variation of the phenomenon and to establish the shape of its fluctuations.

5. Weakness of the estimator of number of boys per hundred girls

The behaviour of the indicator $\Gamma = 100\,M/F$ is not so satisfactory from the point of view of the calculus of probabilities. This is because, unlike $S = M/N$ where the numerator is random and the denominator is known, this time both terms of the ratio are random. There are several possible ways of showing that the dispersal of the random variable realized by Γ will be wider than that of Σ.

Having constructed, for example, a confidence interval for S, we have a maximum amplitude α for that interval and its probability Π. Therefore, the interval can be expressed in its equivalent for Γ.

$$B_1 = \frac{(S-\alpha)}{(1-S+\alpha)} \leq \frac{\Gamma}{100} \leq \frac{(S+\alpha)}{(1-S-\alpha)} = B_2$$

The semi-interval $\frac{1}{2}(B_2 - B_1)$, similar to α, is then:

$$\frac{1}{2}\left[\frac{(S+\alpha)}{(1-S-\alpha)} - \frac{(S-\alpha)}{(1-S+\alpha)}\right] \geq \frac{\alpha}{(1-S)^2} \approx 4\alpha$$

Thus, at the same 95% level, in order to take a decision on a deviation in the order of 0.1% (0.1 of a point respectively), calculation of the proportion of boys among births ($S = M/N$) calls for roughly a million cases, while that of the ratio of boys to girls ($\Gamma = 100\,M/F$) requires 16 times more. From this it is clear that, of the two traditional indicators, the only one through which we can hope to assess variations satisfactorily from the point of view of the calculus of probabilities is the proportion S, and so the number of boys per hundred girls Γ should be abandoned. That is why, in this book, we have systematically used the index that corresponds to the most precise probabilistic estimator: the proportion of one sex out of the total of known cases.

LIST OF REFERENCES

[Anonymous]. (1808). *Mémoire historique et instructif sur l'Hospice de la Maternité.* Paris: Imprimerie des hospices civils.

[Anonymous]. (1825). Résumé spéculatif. *Annales des Sciences Naturelles, V*, pp. 21–29 (with an addition entitled Nota bene, p. 46).

Aberdam, S. (2004). *Démographes et démocrates. L'Oeuvre du comité de division de la Convention nationale. Etude d'histoire révolutionnaire.* Paris: S.E.R.

Allexandre-Lefevre, E. (2002). *L'Infanticide devant le Parlement de Paris de 1750 à 1790.* Ph. D., Univ. Panthéon-Assas.

Alquié, F. (1957). Le Savoir philosophique. *Encyclopédie française*, XIX, Paris: Société nouvelle de l'Encyclopédie Française, pp. 19.04.10–19.04.14.

Arbuthnot, J. (1692) *Of the laws of chance. Or a method of calculation of the hazards of game, Plainly demonstrated and applied to games at present most in use*... London: Motte & Taylor.

Arbuthnot, J. (1710). An argument for divine providence, taken from the constant regularity observ[e]d in the births of both sexes. *Philosophical Transactions.* London, XXVII, pp. 186–190.

Armatte, M. (1995). *Histoire du modèle linéaire. Formes et usages en statistique et économétrie jusqu'en 1945.* Paris: Ph. D., E.H.E.S.S.

Armatte, M. (2004). La Théorie des erreurs (1750–1850). Enjeux, problématiques, résultats. In Barbin & Lamarche, pp. 141–160.

Armatte, M., & Dahan, A. (eds.). (2004). *Modèles et modélisation 1950–2000*, special issue of the *Revue d'histoire des sciences*, 57(2).

Armengaud, A. (1973). L'attitude de la société à l'égard de l'enfant au XIXe siècle. *Annales de démographie historique*, pp. 303–312.

Arokiasamy, P. (2004). Regional patterns of sex bias and excess female child mortality in India. *Population-E*, *59*(6), pp. 833–864.

Attané, I. (2005). *Une Chine sans femmes?* Paris: Perrin.

Attané, I., & Véron, J. (2005). *Gender discriminations among young children in Asia.* Pondichéry: Institut français de Pondichéry & Paris: CEPED.

Avdeev, A., & Monnier, A. (1996). *Mouvement de la population de la Russie 1959–1994. Tableaux démographiques.* Paris: éditions de l'I.N.E.D.

Babbage, C. (1826). *A comparative view of the various institutions for the assurance of lives.* London: Mawman.

Babbage, C. (1829). A Letter to the right Hon. T. P. Courtenay, on the proportionate number of births of the two sexes under different circumstances. *Edinburgh Journal of Science, 1*, pp. 85–104 (art. XIII; dated May 7th, 1829).

Babbage, C., (1864). *Passages from the life of a philosopher*. London: Longman (later ed.: New Brunswick: Rutgers University Press, 1994).

Babbage, C. (1989). *The works of Charles Babbage*. New York: New York University Press, 11 Vols. (ed. Martin Cambell-Kelly).

Bachelard, G. (1936). Le Surrationalisme. In *Inquisitions. Organe du groupe d'études pour la phénoménologie humaine*, (1), pp. 1–6 (reprinted in Henri Béhar, *Du Surréalisme au Front Populaire. Inquisitions*, Paris, éditions du CNRS, 1990).

Bacon, F. (1620). *Novum organum*. In J. Spedding, R. L. Ellis, & D. D. Heath (Trans. English), *The works of Francis Bacon*, Vol. VIII, Boston: Taggard and Thompson, 1863.

Bailly, Dr. (1825). Considérations sur l'influence des circonstances extérieures dans les conceptions et les naissances masculines et féminines. *Annales des Sciences Naturelles, V*, pp. 47 & seq.

Baker, K. M. (1975). *Condorcet, from natural philosophy to social mathematics*. Chicago: University of Chicago Press.

Baker, K. M. (ed.) (1976). *Condorcet: selected writings*. Indianapolis: Bobss-Merrill, 1976.

Banister, J. (2004). Shortage of girls in China today. *Journal of Population Research, 21*(1), pp. 19–45.

Barbin, E., & Lamarche, J.-P. (eds.). (2004). *Histoires de probabilités et de statistiques*. Paris: Ellipses.

Barbut, M. (1984). Sur quelques propriétés élémentaires des fonctions de concentration de C. Gini, *Mathématiques & Sciences Humaines, 22*(88), pp. 5–20.

Baudelot, C., & Establet, R. (1984). *Durkheim et le suicide*. Paris: P.U.F.

Bayes, T. (1764). An Essay towards Solving a Problem in the Doctrine of Chances, *Philosophical Transactions, LIII* (1763), pp. 370–418.

Beaud, J.-P., & Prévost, J.-G. (eds.). (2000). *L'Ère des chiffres, systèmes statistiques et traditions nationales. The Age of Numbers, Statistical Systems and National Traditions*. Sillery: Presses de l'Université du Québec.

Beaumel, C., Désesquelles, A., Richet-Mastain, L., & Vatan, M. (2004). *La Situation démographique en 2002. Mouvement de la population*. Paris: I.N.S.E.E.

Bergson, H. (1932). *Les Deux sources de la morale et de la religion*. Paris: Alcan.

Bergson, H. (1995). *Leçons d'histoire de la philosophie moderne. Théorie de l'âme*. Paris: P.U.F.

Bernard, C. (1865). *Introduction à l'étude de la médecine expérimentale*. Paris: Baillière.

Bernard, C. (1878). *La Science expérimentale*. Paris: Baillière.

Bernoulli, J. (1713). *Ars conjectandi. Opus posthumum*. Basileae: Thurnisii (trans. by E. D. Sylla, *The Art of Conjecturing, Together with Letter to a Friend on Sets in Court Tennis*. Baltimore: Johns Hopkins University Press, 2006).

Bertillon, L.-A. (1876). Mort-nés. In A. Dechambre (Ed.), *Dictionnaire encyclopédique des sciences médicales*, Paris: Masson & Asselin, Vol. 10, pp. 2–28.

Bertrand, J. (1889). *Calcul des probabilités*. Paris: Gauthier-Villars

Besnard, P., Borlandi, M., & Vogt, P. (1993). *Division du travail et lien social. La thèse de Durkheim un siècle après*. Paris: P.U.F.

Bideau, A., & Bardet, J.-P. (1988). La Fécondité. In Dupâquier, pp. 349–398.

Bindra, K., Christensen, J., & Ficenec, S. (2004). *Vital Statistics of California*. Sacramento, State of California: Department of Health Services.

Birg, H. (eds.) (2001), *Demographie and Ethik - das Werk von Johann Peter Süssmilch mit einem Blick auf David Hume and Thomas R. Malthus*, Düsseldorf, Verlag Wirtschaft and Finanzen.

Black, E. (2001). *IBM and the Holocaust: the strategic alliance between Nazi Germany and America's most powerful corporation.* New York: Three Rivers Press.

Blomac, N. de (1991). *La Gloire et le jeu. Des hommes et des chevaux, 1766–1866.* Paris: Fayard.

Boltanski, L. (2004). *La Condition fœtale. Une sociologie de l'engendrement et de l'avortement.* Paris: Gallimard.

Bonneuil, N. (1997). *Transformation of the French demographic landscape, 1806–1906.* Oxford: Oxford University Press.

Borel, É. (1909). *Éléments de la théorie des probabilités.* Paris: Hermann (3rd ed., 1924).

Bouglé, C. (1899). *Les Idées égalitaires. Etude sociologique.* Paris: Alcan (2nd ed., 1908).

Bourdeau, M. (2006). *Les Trois états. Science, théologie et métaphysique chez Auguste Comte.* Paris: Cerf.

Bourdieu, P. (1972). *Esquisse d'une théorie de la pratique.* Genève: Droz.

Bourdieu, P. (1977) Sur le pouvoir symbolique, *Annales E.S.C.*, no. 3, pp. 405–411.

Bourdieu, P. (1980). *Le Sens pratique.* Paris: Minuit.

Bourdieu, P. (1984). *Homo academicus.* Paris: Minuit.

Bourdieu, P. (1998). *La Domination masculine.* Paris: Seuil.

Bourdieu, P. (2001). *Science de la science et réflexivité.* Paris: Raisons d'Agir.

Bourdieu, P. Chamboredon, J.-C., & Passeron, J.-C. (1968). *Le Métier de sociologue.* Paris-La Haye: Mouton.

Bourguet, M.-N. (1989). *Déchiffrer la France. La statistique départementale à l'époque napoléonienne.* Paris: éditions des Archives contemporaines.

Box, G. E. P., & Jenkins, G. M. (1970). *Time series analysis, forecasting and control.* San Francisco: Holden-Day, (2nd ed., 1976, 3rd ed., Upper Saddle River: Prentice Hall, 1994).

Brian, É. (1989). Statistique administrative et internationalisme statistique pendant la seconde moitié du XIXe siècle. *Histoire & Mesure, IV*(3–4), pp. 201–224.

Brian, É. (1991). Les Moyennes à la Société de statistique de Paris (1874–1885). *Moyenne, milieu, centre. Histoires et usages* (eds. J. Feldman, G. Lagneau & B. Matalon), Paris: éditions de l'E.H.E.S.S., pp. 107–134.

Brian, É. (1992). L'Oeil de la science incessamment ouvert. Trois variantes de l'objectivisme statistique. *Communications*, (54), pp. 89–103.

Brian, É. (1994a). *La Mesure de l'État. Administrateurs et géomètres au XVIIIe siècle.* Paris: Albin Michel.

Brian, É. (1994b). La Science administrative du baron de Montyon. In *Moheau. Recherches et considérations sur la population de la France.* Paris: éditions de l'I.N.E.D., pp. 367–382.

Brian, É. (1996a) Calepin. Repérage en vue d'une histoire réflexive de l'objectivation. *Enquête. Anthropologie, Histoire, Sociologie*, (2), pp. 193–222.

Brian, É. (1996b). L'Objet du doute. Les articles de D'Alembert sur l'analyse des hasards dans les quatre premiers tomes de l'*Encyclopédie. Recherches sur Diderot & sur l'Encyclopédie*, (21), pp. 163–178.

Brian, É. (2001a). *Staatsvermessungen. Condorcet, Laplace, Turgot und das Denken der Verwaltung.* Wien: Springer Verlag.

Brian, É. (2001b). Les Objets de la chose. Théorie du hasard et surréalisme au XXe siècle. *Revue de synthèse, 122* (2–4), pp. 473–502.

Brian, É. (2002). Transactions statistiques au XIXe siècle. Mouvements internationaux de capitaux symboliques. *Actes de la recherche en sciences sociales*, Paris, (145), pp. 34–46.

Brian, É. (2006). Les Phénomènes sociaux que saisissait Jakob Bernoulli, aperçus de Condorcet à Auguste Comte. *Electronic Journal for History of Probability and Statistics, 2*(1 *bis*), (www.jehps.net).

Brian, É., & Alunni, C. (eds.). (2001). *Objets d'échelles*, special issue of *Revue de synthèse, 122*(1).
Brian, É., & Jaisson, M. (2005a). Probability, biology and sociology in human sex-ratio at birth. A note on the trace of the First World War. *Electronic Journal for History of Probability and Statistics, 1*(1), (www.jehps.net).
Brian, É., & Jaisson, M. (2005b). Nombre et mémoire. Halbwachs sociologue probabiliste. In H. Krapoth & D. Laborde (eds.), *Erinnerung und Gesellschaft. Mémoire et société*, Wiesbaden: VS Verlag für Sozialwissenschaften, pp. 127–151.
Brian, É., & Jaisson, M (2005c). La Dynamique du sex-ratio à la naissance. Un modèle probabiliste construit sur les cas chinois et français. *Cahiers du CAMS*, no. 248, *Série Histoire du calcul des probabilités et de la statistique*, no. 68.
Brian, É., & Jaisson, M. (2005d). La Filiation physiologique du sex ratio à la naissance. *In M. Halbwachs (2005)*, pp. 85–100.
Brian, É., & Jaisson, M. (2007). *Le Sexisme de la première heure. Hasard et sociologie.* Paris: Raisons d'Agir.
British Association for the Advancement of Science. (1840). *Report of the Ninth Meeting of the British Association for the Advancement of Science Held at Birmingham in August 1839.* London.
Browne, J. (1995, 2002). *Charles Darwin. Voyaging*, Vol. 1. London: Jonathan Cape; *Charles Darwin. The Power of Place*, Vol. 2. London: Jonathan Cape.
Bru, B. (2003). De la Physico-théologie démographique à la physique statistique. In Martin (2003), pp. 71–88.
Bryon, T. J. (1828). *Le Haras français, contenant en langues française et anglaise la généalogie de tous les étalons, juments, etc. célèbres connus dans ce pays [...] – The French Stud-Book [...].* Paris: Didot aîné.
Burdach, C. F. (1835–1840). *Die Physiologie als Erfahrungswissenschaft.* Leipzig: Voss, 6 vols.
Buttay, F. (2005). *'Fortuna'. Usages politiques d'une allégorie morale (Italie et France vers 1460–70 – vers 1600).* Paris: Ph. D., Université Paris-IV.
Cahen, L. (1922). Condorcet inédit. Notes pour le Tableau des progrès de l'esprit humain. *La Révolution française*, (75), pp. 193–212.
Cai, Y., & Lavely, W. (2003). China's missing girls: Numerical estimates and effects on population growth. *China Review*, (3), pp. 13–29.
Campbell-Kelly, M. (1994). Charles Babbage and the assurance of lives. *Annals of the History of Computing, 16*(3), pp. 5–14.
Carnot, L. (1797). *Réflexions sur la métaphysique du calcul infinitésimal.* Paris: Duprat (Paris: Gauthier-Villars, 1921, 2 vols.).
Carvallo, E. (1912). *Le Calcul des probabilités et ses applications.* Paris: Gauthier-Villars.
Caselli, G., Vallin, J., & Wunsch, G. (eds.). (2001, 2002a, 2002b, 2003, 2004a, 2004b, 2006a, 2006b). *Démographie. Analyse et synthèse.* Paris: éditions de l'I.N.E.D., 8 vols.
Catalano, R., Bruckner T., Gould, J., Eskenazi, B., & Anderson, E. (2005). Sex ratios in California following the terrorist attacks of September 11, 2001. *Human Reproduction, 20*(5), pp. 1221–1227.
Catalano, R., & Bruckner T. (2006). Secondary sex ratios and male lifespan: damaged or culled cohorts. *Proceedings of the National Academy of Sciences of the United States of America, 103*(5), pp. 1639–1643.
Cavalli-Sforza, L. L., & Bodmer, W. F. (1999). *The genetics of human populations.* Mineola, N.Y.: Dover Publications.

Cerutti, S., & Pomata, G. (2001). *Fatti: storie dell'evidenza empirica. Quaderni Storici, 108*(3). Bologna.

Chabot, G.-A. (1818). *Commentaire sur la loi des successions formant le titre premier du livre troisième du Code civil* (3 vols.). Paris: Nêve.

Chabrol de Volvic, G. (ed.). (1821, 1823, 1826, 1829). *Recherches statistiques sur la ville de Paris et le département de la Seine... Recueil de tableaux dressés et réunis d'après les ordres de M. le Comte de Chabrol, préfet du département* (4 vols.). Paris, Ballard, then Paris, Imprimerie royale

Chappey, J.-L. (2002). *La Société des observateurs de l'homme (1799–1804). Des anthropologues au temps de Bonaparte*. Paris: S.E.R.

Charbonneau, L. (1993). *Catalogue des manuscrits de Joseph Fourier*. Paris: S.F.H.S.T.

Cifoletti, G. C. (ed.). (2006). *Mathematics and Rhetoric* special issue of *Early Modern Science and Medicine, 11*(4).

Cléro, J.-P. (2004). *Les Raisons de la fiction. Les philosophes et les mathématiques*. Paris: Armand Colin.

Coale, A. J., & Banister, J. (1994). Five decades of missing females in China. *Demography, 31*(3), pp. 459–479.

[Code]. (1841). *Code Napoleon or The French Civil Code. Literally Translated from the Original and Official Edition, Published at Paris, in 1804*. New-York: Halsted & Voorhies.

Comte, A. (1822). Plan des travaux scientifiques nécessaires pour réorganiser la société. *Ecrits de jeunesse (1816–1828) suivis du Mémoire sur la Cosmogonie de Laplace (1835)*. Paris: Mouton, 1970.

Comte, A. (1830). *Cours de philosophie positive. Tome premier contenant les préliminaires généraux et la philosophie mathématique*. Paris: Rouen Frères (later ed. Hermann, 1998).

Comte, A. (1839). *Cours de philosophie positive. Tome quatrième contenant la philosophie sociale et les conclusions générales*. Paris: Bachelier (Hermann, 1998, English trans. by Harriet Martineau, *Cours de philosophie positive. The positive philosophy*, New York, Calvin Blanchard, 1855, reprint: New York, AMS Press, 1974).

Condorcet. (1781). Sur les probabilités., *Histoire de l'Académie royale des sciences pour l'année 1778*. Paris: Imprimerie royale, pp. 43–46, Sur un mémoire de Morand sur la population. *id. pour l'année 1779*, Paris: Imprimerie royale, 1782, pp. 23–27.

Condorcet. (1786). Éloge de M. Euler, Éloge de M. D'Alembert, Éloge de M. Wargentin. *Histoire de l'Académie royale des sciences pour l'année 1783*, Paris: Imprimerie royale, pp. 37–68, 76–120, 128–132.

Condorcet. (1784–1787). Mémoire sur le calcul des probabilités (6 articles in 4 publications). *Mémoires de l'Académie royale des sciences. Années 1781, 1782, 1783, 1784*, Paris, pp. 707–728, 694–691, 539–559, 454–477 (later ed. in Condorcet, 1994, pp. 387–448).

Condorcet. (1791). Éloge de Buffon. *Histoire de l'Académie royale des sciences*, année 1788, Paris, pp. 50–84 (published in: Buffon, *Histoire naturelle*, Paris, an VIII, Vol. 22, pp. 5–64).

Condorcet. (1785). *Essai sur l'application de l'analyse à la probabilité des décisions rendues à la pluralité des voix*. Paris: Imprimerie royale.

Condorcet. (2004). *Esquisse d'un tableau historique des progrès de l'esprit humain. Ouvrage posthume de Condorcet*. Paris: Agasse, An III (recent critical ed., Condorcet, 2004, pp. 228–470, first English trans.: *Outlines of an Historical View of the Progress of the Human Mind*. London: J. Johnson, 1795).

Condorcet. (1994). *Arithmétique politique. Textes rares ou inédits* (eds. B. Bru & P. Crépel). Paris: éditions de l'I.N.E.D.

Condorcet. (2004). *Tableau historique de l'esprit humain. Projets, Esquisse, Fragments et Notes (1772–1794)*. Paris: éditions de l'I.N.E.D.

Conry, Y. (1981). Organisme et organisation. De Darwin à la génétique des populations. *Revue de synthèse, 103–104.*, pp. 291–221.

Coumet, E. (1970). La Théorie du hasard est-elle née par hasard? *Annales E.S.C.*, (3), pp. 574–598.

Coumet, E. (2003). Auguste Comte. Le calcul des chances, aberration radicale de l'esprit mathématique. *Mathématiques & Sciences Humaines, 41*(162), pp. 9–17.

Courgeau, D. (2007). *Multilevel Synthesis. From the Group to the Individual.* Dordrecht: Springer.

Cournot, A. (1843). *Exposition de la théorie des chances et des probabilités.* Paris: Hachette.

Crampe-Casnabet, M. (1985). Michèle. *Condorcet, lecteur des Lumières.* Paris: P.U.F.

Cuvier, G. (1831). Analyse des travaux de l'Académie royale des sciences pendant l'année 1827. Partie physique. *Mémoires de l'Académie des sciences de l'Institut de France.* Paris: Firmin-Didot, Vol. 10, pp. CIII–CXC.

D'Alembert, J. Le Rond. (1747). *Réflexions sur la cause générale des vents....* Paris: David l'aîné.

D'Alembert, J. Le Rond. (1754). Croix ou pile. *Encyclopédie*, Vol. IV, pp. 512–513, Déduction. *ibid.*, Vol. IV, 1754, pp. 729–730, Fortuit. *ibid.*, Vol. VII, 1757, pp. 204–205.

D'Alembert, J. Le Rond. (1767). Éclaircissemens sur différens endroits des *Éléments de philosophie. Mélanges de littérature, d'histoire et de philosophie.* Amsterdam: Chatelain, Vol. 5 (*Essai sur les Éléments de philosophie*, Paris: Fayard, 1986, pp. 220–259).

Daguet, F. (2002). *Un Siècle de fécondité française. Caractéristiques et évolution de la fécondité de 1901 à 1999.* Paris: I.N.S.E.E.

Darwin, C. (1859). *On the Origin of Species by Means of Natural Selection, or the Preservation of Favoured Races in the Struggle for Life.* London: John Murray (6th ed. 1872).

Darwin, C. (1868). *The Variation of Animals and Plants under Domestication.* London: John Murray.

Darwin, C. (1871). *The Descent of Man and Selection in Relation to Sex.* London: John Murray (2nd edition, revised and augmented, London: John Murray, 1874, Dutch trans. by Hermanus Hartogh Heys van Zouteveen, *De afstamming van den mensch, en de seksueele teeltkeus*, Delft, Ykema & Van Gyn, 1871–1872, reference French trans., *La Filiation de l'homme et la sélection liée au sexe*, Paris: éditions Syllepse, 2000).

Daston, L. (1988). *Classical Probability in the Enlightenment.* Princeton: Princeton University Press.

Delaporte, F., & Delaporte, C. (2004). Fécondation artificielle. *Dictionnaire de la pensée médicale*, Paris: P.U.F., pp. 481–488.

Demeulenaere-Douyère, C., & Brian, É. (eds.). (2002). *Règlement, usages et science dans la France de l'Absolutisme.* Paris: Lavoisier.

De Morgan, A. (1838). *An Essay on Probabilities, and on their Application to Life Contingencies and Insurance Offices.* London: Longman (partial reprint in Jenkins & Yoneyama, 2000, Vol. 5).

Derham, W. (1713). *Physico-Theology, or a Demonstration of the Being and Attributes of God, from His Works of Creation....* London (later *Physico-theology...being the substance of XVI sermons*, London, 1714).

Dhombres, J., & Robert, J.-B. (1998). *Fourier. Créateur de la physique mathématique.* Paris: Belin.

Dionis, P. (1718). *Traité général des accouchements, qui instruit de tout ce qu'il faut faire pour être habile accoucheur.* Paris: Houry.

Dupâquier, J., & Dupâquier, M. (1985). *Histoire de la démographie. La statistique de la population des origines à 1914.* Paris: Perrin.

Dupâquier, J. (ed.). (1988). *Histoire de la population française. 2. De la Renaissance à 1789.* Paris: P.U.F.

Durand-Richard, M.-J. (1996). L'École algébrique anglaise. Les conditions conceptuelles et institutionnelles d'un calcul symbolique comme fondement de la connaissance., In C. Goldstein, J. Gray & J. Ritter (eds.), *Mathematical Europe*, Paris: éditions de la Maison des sciences de l'homme, pp. 445–477.

Durand-Richard, M.-J. (2001). Révolution industrielle. Logique et signification de l'opératoire. *Revue de synthèse, 122*(2–4), pp. 319–346.

Durkheim, É. (1893). *De la Division du travail social.* Paris: Alcan (P.U.F., 1978, trans. by G. Simpson, *The Division of Labor in Society*, Glencoe Illinois: Free Press, 1933).

Durkheim, É. (1895). *Les Règles de la méthode sociologique.* Paris: Alcan (P.U.F., 1981, trans. by S. A. Solovay & J. H. Mueller, *The Rules of Sociological Method*, Glencoe Illinois: Free Press, 1950).

Durkheim, É. (1897). *Le Suicide.* Paris: Alcan, (P.U.F., 1967, trans. by J. A. Spaulding & G. Simpson, *Suicide, a Study in Sociology*, New York: Free Press, 1951).

Durkheim, É. (1912). *Les Formes élémentaires de la vie religieuse.* Paris: Alcan, (P.U.F., 1979, trans. by J. W. Swain, *The Elementary Forms of the Religious Life*, New York: The Free Press, 1965).

Düsing, C. (1883). *Die Faktoren, welche die Sexualität entscheiden.* Jena: Fischer.

Düsing, C. (1884). *Die Regulierung das Geschlechtsverhältnisses bei der Vermehrung der Menschen, Tiere und Pflanzen.* Jena: Fischer.

Düsing, C. (1890). *Das Geschlechtsverhältnis der Geburten in Preussen.* Jena: Fischer.

Edgeworth, F. Y. (1892). [Review of] Das Geschlechtsverhältnis der Geburten in Preussen von Dr. C. Düsing, Jena, Gustav Fischer [1890]. In *The Economic Journal*, Vol. 2(6), pp. 337–340.

Edwards, A. W. F. (1998). Natural selection and the sex ratio. Fisher's sources. *American Naturalist*, (151), pp. 564–569.

Edwards, A. W. F. (2000). Carl Düsing on the regulation of the sex ratio., *Theoretical Population Biology*, (58), pp. 255–257.

Ehrard, J. (1963). *L'Idée de nature en France dans la première moitié du XVIIIe siècle.* Paris: SEVPEN (later ed.: Albin Michel, 1994).

Enros, P. C. (1979). *The Analytical Society: Mathematics at Cambridge University in the Early Nineteenth Century.* Toronto: Ph. D., University of Toronto.

Ericsson, R. J., & Ericsson, S. A. (1999). Sex ratios. In Knobil & Neill, Vol. 4, pp. 431–437.

Fagot-Largeault, A. (1989). *Les Causes de la mort. Histoire naturelle et facteurs de risque.* Paris: Vrin.

Febvre, L. (1942). *Le Problème de l'incroyance au XVIe siècle. La religion de Rabelais.* Paris: Albin Michel.

Feller, W. (1950). *An Introduction to Probability Theory and Its Applications.* New York: Wiley, 2 vols. (3rd ed., 1967–1968).

Ferdinand, U. (1997). *Das Malthusische Erbe. Entwicklungsstränge der Bevölkerungstheorie im 19. Jahrhundert und deren Einfluß auf die radikale Frauenbewegung in Deutschland.* Berlin: Ph. D., Freie Universität Berlin.

Fisher, R. A. (1925). *Statistical Methods for Research Workers*. Edinburgh: Oliver and Boyd (later *Statistical Methods, Experimental Design, and Scientific Inference*, Oxford & New York: Oxford University Press, 1990).

Fisher, R. A. (1926). The arrangement of field experiments. *Journal of the Ministry of Agriculture of Great Britain, 33*, pp. 503–513.

Fisher, R. A. (1930). *The Genetical Theory of Natural Selection*. Oxford: Clarendon Press (2nd ed., New York, Dover, 1958).

Fisher, R. A. (1935). *The Design of Experiments*. Edinburgh: Oliver & Boyd.

Fisher, R. A., & Wishart, J. (1930). The arrangement of field experiments and the statistical reduction of the result. *Imperial Bureau of Soil Science*, Technical communication no. 10, London: HMSO.

Fourier, J. (1821). Notions générales sur la population. In Chabrol de Volvic (1821), pp. IX–XLVIII.

Fourier, J. (1823). Mémoire sur la population de la Ville de Paris depuis la fin du XVIIe siècle In Chabrol de Volvic (1823), pp. XIII–XXVIII.

Fourier, J. (1824). Règle usuelle pour la recherche des résultats moyens d'un grand nombre d'observations. *Bulletin des sciences mathématiques, physique et chimique*, (2), pp. 88–90.

Fourier, J. (1826). Mémoire sur les résultats moyens déduits d'un grand nombre d'observations. In Chabrol de Volvic (1826), pp. IX–XXXI (later in *Œuvres de Fourier*, Paris: Gauthier-Villars, 1890, Vol. II, pp. 525–545).

Fourier, J. (1829). Second mémoire sur les résultats moyens et sur les erreurs des mesures. In Chabrol de Volvic (1829) pp. IX–XLVIII (later in *Œuvres de Fourier*, Paris: Gauthier-Villars, 1890, Vol. II, pp. 549–590).

Fréchet, M., & Halbwachs, M. (1924). *Le Calcul des probabilités à la portée de tous*. Paris: Dunod.

Freundlich, F. (1989). *Le Monde du jeu à Paris au XVIIIe siècle*. Paris: Ph. D., Université Paris-I.

Gaillard, N. (2000). *L'Infanticide en cour d'assises au XXe siècle*. Rennes Med. D., Université de Rennes-I.

Galton, F. (1869). *Hereditary genius. An inquiry into its laws and consequences*. London: Macmillan.

Galton, F. (1886). Regression towards mediocrity in hereditary stature. *Journal of the Anthropological Institute, 15*, pp. 246–263.

Gayon, J. (1992). *Darwin et l'après-Darwin: une histoire de l'hypothèse de sélection naturelle*. Paris: Kimé (trans. as *Darwinism's Struggle for Survival: Heredity and the Hypothesis of Natural Selection*, Cambridge: Cambridge University Press, 1998).

Gayon, J. (1996). Fisher, Ronald Aylmer. In Tort (1996), pp. 1671–1674.

Gillispie, C. C. (1980). *Science and Polity in France at the End of the Old Regime*. Princeton: Princeton University Press.

Gillispie, C. C. et al. (1997). *Pierre-Simon Laplace, 1749–1827. A Life in Exact Science*. Princeton: Princeton University Press.

Gillispie, C. C. (2004). *Science and Polity in France. The Revolutionary and Napoleonic Years*. Princeton: Princeton University Press.

Gini, C. (1908). *Il sesso dal punto di vista statistico. Le leggi della produzione dei sessi*. Milano-Palermo-Napoli: Biblioteca Sandron di scienze e lettere.

Girardin, J. (1845). *Notices biographiques sur de Morel-Vindé, d'Arcet et Mathieu de Dombasle*. Rouen: Péron.

Girou de Buzareingues, C. (1825). Observations sur les rapports de la mère et du père avec les produits, relativement au sexe et à la ressemblance. *Annales des sciences naturelles, V*, pp. 30–46.

Girou de Buzareingues, C. (1828). *De la Génération. Suivi d'un Mémoire sur la distribution et le rapport des sexes en France*. Paris: Huzard.

Goffman, E. (1977). The arrangement between the sexes. *Theory and Society*, *4*(3), pp. 301–331.

Goncourt, E. de, & Goncourt, J. de (2004). *Journal. Mémoires de la vie littéraire* (3 Vols.). Paris: Robert Laffont.

Gouhier, H. (1931). *La Vie d'Auguste Comte*. Paris: Gallimard.

Gouhier, H. (1933–1941). *La Jeunesse d'Auguste Comte et la formation du positivisme*. Paris: Vrin.

Grattan-Guinness, I. (ed.). (2005). *Landmark Writings in Western Mathematics 1640–1940*. Amsterdam: Elsevier.

Grattan-Guinness, I., & Ravetz, J. R. (eds.). (1972). *Joseph Fourier 1768–1830. A Survey of his Life and Work*. Cambridge: MIT Press.

Graunt, J. (1662). *Natural and Political Observations* [...] *made upon the Bills of Mortality*.... London: T. Raycroft (reprint: New York: Arno Press, 1975, trans. by É. Vilquin, *Observations naturelles et politiques ... faites sur les bulletins de mortalité* .... Paris: éditions de l'I.N.E.D., 1977).

Grech, V., Vassallo-Agius, P., & Savona-Ventura, C. (2003). Secular trends in sex ratios at birth in North America and Europe over the second half of the 20th century. *Journal of Epidemiology and Community Health, 57*, pp. 612–615.

Grenier, J.-Y. (1996). *L'Économie d'Ancien Régime. Un monde de l'échange et de l'incertitude*. Paris: Albin Michel.

Guégan, I. (1991). *Inventaire des enquêtes administratives et statistiques. 1789–1795*. Paris: CTHS-CHRF.

Guilbaud, G. Th. (1952). Les Théories de l'intérêt général et le problème logique de l'agrégation., *Economie appliquée, V*(4), pp. 501–584.

Guilhaumou, J. (2002). *Sieyès et l'ordre de la langue. L'invention de la politique moderne*. Paris: Kimé.

Guilmoto, C. Z. (2005). Spatial regression and determinants of juvenile sex ratio in India., *Female Deficit in Asia, Trends and Perspectives*, Singapore (posted as www.cicred.org/Eng/Seminars/FDA/papers/68_Guilmoto.pdf).

Gutberger, J. (1994). *Volk, Raum und Sozialstruktur: Sozialstruktur- und Sozialraumforschung im Dritten Reich*. Goetti Ph. D., University of Goettingen.

Hacking, I. (1975). *The emergence of probability: A philosophical study of early ideas about probability, induction and statistical inference*. Cambridge: Cambridge University Press.

Hacking, I. (1990). *The taming of chance*. Cambridge: Cambridge, University Press.

Hacking, I. (2001). *An introduction to probability and inductive logic*. Cambridge: Cambridge University Press.

Hahn, R. (2004). *Le Système du monde. Pierre Simon Laplace, un itinéraire dans la science*, Paris, Gallimard.

Halbwachs, M. (1906). *Leibniz*. Paris: Delaplane (later eds.: Mellottée, 1928 and 1950).

Halbwachs, M. (1909). *Les Expropriations et le prix des terrains à Paris (1860–1900)*. Paris: Rieder-Cornély (revised ed.: *La Population et les tracés de voies à Paris depuis cent ans*, Paris, P.U.F., 1928).

Halbwachs, M. (1912). *La Théorie de l'homme moyen. Essai sur Quetelet et la statistique morale*. Paris: Alcan.

Halbwachs, M. (1913). On Gini, 1908, *L'Année sociologique 1909–1912*, XII, pp. 826–830.

Halbwachs, M. (1923). L'Expérimentation statistique et les probabilités, *Revue philosophique, 96*, pp. 340–371 (later ed.: *Classes sociales...*, 1972, pp. 275–307).

Halbwachs, M. (1925). *Les Cadres sociaux de la mémoire*. Paris: Alcan (later ed.: Albin Michel, 1994).

Halbwachs, M. (1930). *Les Causes du suicide*. Paris: Alcan (later ed.: P.U.F., 2002).

Halbwachs, M. (1931). La Statistique et les sciences sociales en France. In J. Čada (ed.), *La France d'aujourd'hui...*, Prague: Spolek posluchačů komerčního inženýrství, pp. 272–286 (later ed. in: *Le Point de vue du nombre...*, 2005, pp. 369–380).

Halbwachs, M. (1933). Recherches statistiques sur la détermination du sexe à la naissance. *Journal de la Société de statistique de Paris, 74*(5), mai, pp. 164–195 (later ed. in: *Le Point de vue du nombre...*, 2005, pp. 381–404).

Halbwachs, M. (1935). Les Facteurs biologiques de la population. *Revue philosophique, 119*, pp. 285–303.

Halbwachs, M. (1937). Note sur l'emploi des coefficients de corrélation ou de covariation principalement en sociologie. *Annales sociologiques, C*(2), pp. 141–144.

Halbwachs, M. (1938). *Morphologie sociale*. Paris: Armand Colin (later ed.: 1970).

Halbwachs, M. (1941). *La Topographie légendaire des évangiles en Terre Sainte. Etude de mémoire collective*. Paris: P.U.F. (later ed.: 1971).

Halbwachs, M. (1950). *La Mémoire collective*. Paris: P.U.F. (later and revised ed.: Albin Michel, 1997).

Halbwachs, M. (1972). *Classes sociales et morphologie* (V. Karady ed.). Paris: Minuit.

Halbwachs, M. (et al.) (2005). *Le Point de vue du nombre (1936)*. Paris: éditions de l'I.N.E.D. (M. Jaisson & É. Brian eds., first publ. in *Encyclopédie française*, Vol. VII, Paris: Comité de l'Encyclopédie Française, 1936, pp. 7.76.01–7.94.04).

Hald, A. (1998). *A history of mathematical statistics from 1750 to 1930*. New York: Wiley.

Hald, A. (1990). *A history of probability and statistics, and their applications before 1750*. New York, Wiley.

Hamilton, W. D. (1967). Extraordinary Sex Ratios, *Science*, (156), pp. 477–488.

Hamon, É. (2001). Évaluer la peine des hommes. La mesure de l'architecture à Paris à la fin du Moyen Âge (XVe-début XVIe siècles). *Histoire & mesure, XVI*(3/4), pp. 283–307.

Hardy, I. C. W. (ed.). (2002). *Sex Ratios. Concepts and Research Methods*. Cambridge: Cambridge University Press.

Heilbron, J. (1990). *Het ontstaan van de sociologie*. Amsterdam: Prometheus (trans. as *The Rise of Social Theory*. Cambridge: Polity Press, 1995).

Heilbron, J., Magnusson, L., & Wittrock, B. (eds.). (1998). *The rise of social sciences and the formation of modernity. Sociology of the sciences: An international yearbook*. Dordrecht: Kluwer Academic Publications.

Héritier, F. (1996, 2002). *Masculin-féminin. La pensée de la différence*. Paris: Odile Jacob, *Masculin-féminin. II, Dissoudre la hiérarchie*. Paris: Odile Jacob.

Heuschling, X. (1882a). La Statistique internationale, son origine et ses progrès. *Journal de la Société de statistique de Paris*, XXIII, pp. 287–295.

Heuschling, X. (1882b). Vingt-cinquième anniversaire du Congrès international de statistique (1853–1878). *Société de statistique de Paris, son histoire, ses travaux, son personnel*, Nancy:, Berger-Levrault, pp. 102–112.

Heyde, C., & Seneta, E. (eds.). (2001). *Statisticians of the Centuries*. New York & Berlin: Springer Verlag.

Hofacker, J. D. (1827). Aus einem Schreiben des hn. Professors Hofacker in Tübingen an den herausgeber des med. chir. Zeitung. *Medicinisch-chirurgische Zeitung*, (100), pp. 398–399.

Hoffmann, J. G. (1842). *Zur Judenfrage: statistische Erörterung. Anzahl... der Juden im Preussischem Staate nach einer Vergleichung der Zählungen im Jahren 1840 und 1822*. Berlin: Lesekabinet.

Hoffmann, J. G. (1844). Übersicht der bei dem statistischen Bureau zu Berlin vorhandenen Nachrichten über die Anzahl und Vermehrung der Juden im preussischen Staate. *Philologische und historische Abhandlungen der königlichen Akademie der Wissenschaften zu Berlin. Aus dem Jahr 1842*, pp. 224–243.

Houston, J. W. (2003). *The legend of fire horse woman*. New York: Kensington Books.

Hume, D. (1748). *Philosophical essays concerning human understanding*. London: A. Millar (later: *An Enquiry concerning Human Understanding*, La Salle (Illinois). Open Court, 1988).

I.N.S.E.E. [Institut national de la statistique et des études économiques]. (1948, 1961, 1990, 2001, 2004). *Statistique du mouvement de la population. Première partie. Mariages, divorces, naissances, décès*, Vol. 22, année 1945, Paris: P.U.F., *Annuaire statistique de la France – Rétrospectif*, Vol. 66. Paris, P.U.F., *Annuaire rétrospectif de la France 1948–1988. Séries longues*. Paris, I.N.S.E.E., *La Situation démographique en 1998. Mouvement de la population*. Paris, I.N.S.E.E, *La Situation démographique en 2002. Mouvement de la population*. Paris, I.N.S.E.E.

[International Population Congress, 1935]. (1936). *Bevölkerungsfragen. Bericht des Internationalen Kongresses für Bevölkerungs-wissenschaft. Berlin, 26 August – 1 September 1935*. München: Lehmans.

[International Population Congress, 1937]. (1938). *Congrès international de la population. Paris 1937*. Paris: Hermann, 8 vols. (Actualités scientifiques et industrielles, no. 710–717).

[International Statistical Congress, 1853]. (1853). *Compte rendu des travaux du Congrès général de statistique réuni à Bruxelles les 19, 20, 21 et 22 septembre 1853*. Bruxelles: Hayez.

[International Statistical Congress, 1869]. (1869). *Congrès international de statistique à La Haye. Septième session du 6 au 11 septembre 1869*. La Haye: Martinus Nijhoff, 3 vols. (in particular J. A. Boogaard & L. J. Egeling, La Question des morts-nés dans ses rapports avec le mouvement de la population, Vol. 1, *Programme*, pp. 32–34; Première séance du Mercredi 8 septembre, Vol. 2, pp. 71–81 and Seconde séance du Mercredi 8 septembre, Vol. 2, pp. 81–92).

Ipsen, C. (1996). *Dictating demography. The problem of population in fascist Italy*. Cambridge, Cambridge University Press.

Israel, G., & Nastasi, P. (1998). *Scienza e razza nell'Italia fascista*. Bologna: Il Mulino.

Jenkin, F. (1867). [Review of Darwin's] 'The origin of species'. *The North British Review*, (46), pp. 277–318.

Jenkins, D., & Yoneyama, T. (2000). *The history of insurance* (8 Vols.). London: Pickering & Chatto.

Kaku, K.(1972). Are Physicians Sympathetic to Superstition? A Study of Hinoe-Uma. *Social Biology, 19*(1), pp. 60–64.

Kallenberg, O. (1997). *Foundations of modern probability*. New York: Springer-Verlag.

Keval, S. (1999). *Die schwierige Erinnerung: deutsche Widerstandskämpfer über die Verfolgung und Vernichtung der Juden*. Frankfurt/Main: Campus-Verlag.

Kim, D. S. (2004). Missing girls in South Korea: Trends, levels and regional variations. *Population-E, 59*(6), pp. 865–878.

Korherr, R. (1942). *Zusammenhang zwischen Altersabstand der Eltern und dem Geschlecht der Kinder*, report to Heinrich Himmler, typewritten., Bundesarchiv, Berlin, SS-O Richard Korherr, 201-A/698–710, publ. in French trans. by É. Brian as Mémoire sur la relation entre l'écart d'âge des parents et le sexe des enfants (21 mai 1942), in Halbwachs et al. (2005), pp. 156–163.

Koyré, A. (1957). *From the closed world to the infinite universe*. Baltimore: Johns Hopkins Press.

Krüger, L. et al. (1987, 1989). *The probabilistic revolution*. Cambridge: M.I.T. Press.

Lacroix, S. F. (1816). *An Elementary Treatise on the Differential and Integral Calculus.... Translated from the French*. Cambridge: Deighton, (trans. from *Traité Élémentaire de Calcul Différentiel et de Calcul Intégral. Précédé de réflexions sur la manière d'enseigner les Mathématiques, et d'apprécier dans les examens le savoir de ceux qui les ont étudiées*. Paris: Duprat, 1802).

La Marche, M. du Tertre de (1677). *Instruction familière et utile aux sages-femmes...* (1st ed.). Paris: Houry, 1710.

Lafosse, P.-E. (1775). *Dictionnaire raisonné d'hippiatrique, cavalerie, manège et maréchallerie* [*sic*]. Paris: Boudet, *II*, pp. 340–382.

Lagrée, J. (1991). *La Religion naturelle*. Paris: P.U.F.

Landucci, G. (1996). Darwinisme italien. In Tort, pp. 954–1041.

Laplace, P. S. (1774). Mémoire sur la probabilité des causes par les évènements. *Mémoires [présentés] par divers savants*, Paris: Imprimerie royale, *VI*, pp. 621–656.

Laplace, P. S. (1781). Mémoire sur les probabilités. *Mémoires de l'Académie royale des sciences. Année 1778*. Paris: Imprimerie royale, pp. 227–332 (later in *Œuvres*, Vol. IX, Paris, Gauthier-Villars, 1893, pp. 383–485).

Laplace, P. S. (1785). Mémoire sur les approximations des formules qui sont fonctions de très grand nombres. *Mémoires de l'Académie royale des sciences. Année 1782*. Paris: Imprimerie royale, pp. 1–88 (later in *Œuvres*, Vol. X, Paris, Gauthier-Villars, 1894, pp. 209–291), Suite du mémoire sur les approximations des formules qui sont fonctions de très grand nombres, *ibid., 1783*, Paris, 1786*a*, pp. 423–467 (later in *Œuvres, ibid.*, pp. 295–338).

Laplace, P. S. (1786b). Sur les naissances, les mariages et les morts à Paris, depuis 1771 jusqu'en 1784, et dans toute l'étendue de la France, pendant les années 1781 et 1782. *Mémoires de l'Académie royale des sciences. Année 1783*, Paris: Imprimerie royale, pp. 693–702 (later in *Œuvres*, Vol. XI, Paris, Gauthier-Villars, 1895, pp. 35–46).

Laplace, P. S. (1812). *Théorie analytique des probabilités*. Paris, (reference ed.: 1820; later in *Œuvres*, Vol. VII, Paris, Gauthier-Villars, 1886).

Laplace, P. S. (1814). *Essai philosophique sur les probabilités*. Paris, (reference ed.: 1825; later Paris, Gauthier-Villars, 1921, 2 vols., trans. by F. W. Truscott & F. L. Emory, *A Philosophical Essay on Probabilities*, 1951).

Laplace, P. S. (1992). *Leçons de mathématiques de l'Ecole normale de l'An III*, 1795. Paris: Dunod, pp. 9–167 (Jean Dhombres, ed.).

Léauté, J. (1968). Introduction to *Travaux de l'Institut de sciences criminelles et pénitentiaires, Recherches sur l'infanticide, 1955–1965*. Paris: Dalloz (*Annales de la Faculté de droit et des sciences politiques et économiques de Strasbourg*, Vol. 17).

Lécuyer B.-P. (1977a). Démographie, statistique et hygiène publique sous la monarchie censitaire. *Annales de démographie historique*, Paris, pp. 215–245

Lécuyer, B.-P. (1977b). Médecins et observateurs sociaux. Les *Annales d'hygiène publique et de médecine légale* (1829–1850). *Pour une histoire de la statistique*, Paris: I.N.S.E.E, pp. 445–475.

Legouvé, E. (1849). *Histoire morale des femmes*. Paris: G. Sandré.

Legoyt, A. (1864, 1870). *La France et l'Étranger. Études de statistique comparée* (2 vols.). Paris: Berger-Levrault.

Legoyt, A. (1881). *Le Suicide ancien et moderne, étude historique, philosophique et statistique*. Paris: Drouin.

Leibniz, G. W. (1765). *Nouveaux essais sur l'entendement humain* (1704), later in *Oeuvres philosophiques latines et françoises de feu Mr de Leibnitz*, Amsterdam & Leipzig: Schreuder (trans. by P. Remnant & J. Bennett, *New Essays on Human Understanding*, Cambridge & New York: Cambridge University Press, 1981).

Lepenies, W. (1985). *Die drei Kulturen. Soziologie zwischen Literatur und Wissenschaft*. Munich: Henser (trans. by R. J. Hollingdale, *Between literature and science. The rise of sociology*. Cambridge & New York: Cambridge University Press, 1988).

Lepenies, W. (Ed.). (2003). *Entangled histories and negotiated universals. Centers and peripheries in a changing world*. Frankfurt & New York: Campus.

Lexis, W. (1903). Das Geschlechtsverhältnis der Geborenen und die Wahrscheinlichkeitsrechnung. (1876), In *Abhandlungen zur Theorie der Bevolkerungs- und Moralstatistik*, Jena: Fischer.

Lotka, A. J. (1934, 1939). *Théorie analytique des associations biologiques*, I. *Principes*; II. *Analyse démographique avec application particulière à l'espèce humaine* (2 vols.). Paris: Hermann.

Loty, L. (1995). *La Genèse de l'optimisme et du pessimisme. De Pierre Bayle à la Révolution Française*. Tours: Ph. D., Université de Tours.

Lummaa, V., Merilä, J., & Kause, A. (1998). Adaptative sex ratio variation in pre-industrial human (*Homo sapiens*) populations? *Proceedings of the Royal Society of London, 265*, pp. 563–568.

Mackensen, R. (Ed.). (2002). *Bevölkerungslehre und Bevölkerungspolitik vor 1930*. Opladen: Verlag für Sozialwissenschaften.

Mackensen, R. (Ed.). (2004). *Bevölkerungslehre und Bevölkerungspolitik im Dritten Reich*. Opladen: Verlag für Sozialwissenschaften.

Majerus, M. E. N. (2003). *Sex wars. Genes, bacteria and biased sex ratios*. Princeton: Princeton University Press.

Malinvaud, E. (1955). Relations entre la composition des familles et le taux de masculinité. *Journal de la société de statistique de Paris*, no. 1–3, pp. 49–64.

Malthus, T. (1798). *An essay on the principle of population, as it affects the future improvement of society, with remarks on the speculations of Mr Godwin, M. Condorcet and other Writers*. London: J. Johnson (2nd ed. revised, 1803).

Marc, C.-C.-H. (1829). Extrait d'une lettre du professeur Hofacker au rédacteur de la Gazette médico-chirurgicale d'Innsbruck (autrefois de Salzbourg). *Annales d'hygiène publique et de médecine légale, 1*, 2nd part, pp. 557–558.

Margairaz, D. (2005). *François de Neufchâteau. Biographie intellectuelle*. Paris: Publications de la Sorbonne.

Martin, T. (ed.). (2003). *Arithmétique politique dans la France du XVIIIe siècle*. Paris: éditions de l'I.N.E.D.

Martin, J. A., Hamilton, B. E., Sutton, P. D., Ventura, S. J., Menacker, F., & Munson, M. L. (2005). Births: Final Data for 2003. *National Vital Statistics Reports, 54*(2).

Masuy-Stroobant, G. (2002). Les Déterminants de la santé et de la mortalité infantile. In Caselli, Vallin & Wunsch (2002b), pp. 129–144.

Mathews, T. J., & Hamilton, B. E. (2005). Trend analysis of the sex ratio at birth in the United States. *National Vital Statistics Reports, 53*(20).

Mauss, M., & Fauconnet, P. (1969). Sociologie (1901). In Marcel Mauss, *Œuvres*, Paris: Minuit, Vol. 3, pp. 139–177.

McClellan, J. (2003). *Specialist control. The Publications Committee of the Academie Royale Des Sciences (Paris), 1700–1793*. Philadelphia: American Philosophical Society.

Mercier, L. S. (1994). *Tableau de Paris*. Paris: Mercure de France, Vol. 1.

Memmi, D. (2003). *Faire vivre et laisser mourir. Le gouvernment contemporain de la naissance et de la mort,* Paris, éd. la Découverte.

Métivier, M., Costabel, P., & Dugac, P. (eds.). (1981). *Siméon-Denis Poisson et la science de son temps*. Palaiseau: École polytechnique.

Meusnier, N. (1987). *Jakob Bernoulli. Ars conjectandi. 4ème partie.* Rouen: I.R.E.M. -université de Rouen.

Mises, R. von (1932). Théorie des probabilités. Fondements et applications. *Annales de l'Institut Henri-Poincaré, 3*(2), pp. 137–190.

Moheau, J.-B. (1778). *Recherches et considérations sur la population de la France*. Paris (later ed.: Paris: éditions de l'I.N.E.D., 1994).

Moore, J., & Desmond, A. (2004). Introduction. In *The descent of man, and selection in relation to sex*. London: Penguin Classics.

Moreau, P.-F. (ed.). (2001). *Le Scepticisme au XVIe et au XVIIe siècle*. Paris: Albin Michel (*Le Retour des philosophes antiques à l'Âge classique*, Vol. 2).

Morel-Vindé, C. G. T. de (1813–1816). *Observations sur la monte et l'agnelage*. Paris: Huzard, *Suite des Observations sur la monte et l'agnelage*. Paris: Huzard, *Seconde suite des Observations sur la monte et l'agnelage*. Paris: Huzard, *Troisième et dernière suite des Observations sur la monte et sur l'agnelage, suivie de quelques considérations générales sur les encouragemens [sic] à donner à l'agriculture*. Paris: Huzard.

Neumann-Spallart, F.-X. von (1886). La Fondation de l'Institut international de statistique, aperçu historique. *Bulletin de l'Institut international de statistique, 1*, (1), pp. 1–29.

Newton, I. (1713). *Philosophiæ naturalis principia mathematica. Editio secunda auctior et emendatior*. Cantabrigiæ: Cornelius Crownfield (trans. by A. Motte, *The Mathematical Principles of Natural Philosophy*. London: Benjamin Motte, 1729).

Norton, B. (1981). La Situation intellectuelle du moment des débuts de Fisher en génétique des populations. *Revue de synthèse, 103–104,* pp. 231–246.

Notter, F. (1827). *De qualitatibus parentum in sobolem transeuntibus, praesertim ratione rei equariae*. Tubingae: Osiander (German trans. published under the name of Johann Daniel Hofacker, *Ueber die Eigenschaften welche sich bei Menschen und Thieren von den Eltern auf die Nachkommen vererben, mit besonderer Rücksicht auf die Pferdezucht*. Tübingen: Osiander, 1828).

Ozouf-Marignier, M.-V. (1989). *La Formation des départements. La représentation du territoire français à la fin du 18e siècle*. Paris: éditions de l'E.H.E.S.S.

Ozouf-Marignier, M.-V. (1997). Administration, statistique, aménagement du territoire: l'itinéraire du préfet Chabrol de Volvic (1773–1843). *Revue d'histoire moderne et contemporaine, 44*(1), pp. 19–39.

Panofsky, E. (1927a). Die Perspective als, symbolische Form. In *Vorträge der Bibliothek Warburg, 1924–1925.* Leipzig/ Berlin, Teubner (trans. as *Perspective as Symbolic Form*, New York: Zone Books, 1991).

Panofsky, E. (1927b). Zum Problem der historischen Zeit, publ. in *Aufsätze zu Grundfragen der Kunstwissenschaft*, Berlin: Wissenschaftsverlag Volker Spiess, 1992, pp. 77–83.

Pansini, V. (2002). *L'Oeil du topographe et la science de la guerre. Travail scientifique et perception militaire (1760–1820).* Paris: Ph. D., E.H.E.S.S.

Pearson, K. (1928). Biometry and chronology. *Biometrika, 20* A, pp. 241–252.

Pearson, K. (1978). *The history of statistics in the 17th and 18th centuries, against the changing background of intellectual, scientific and religious thought. Lectures by Karl Pearson given at University College London during the academic sessions 1921–1933.* London: Charles Griffin.

Pergament, E., Todydemir, P. B., & Fiddler, M. (2002). Sex ratio: a biological perspective of 'Sex and the City'. *Reproductive BioMedicine Online, 5*(1), pp. 43–46.

Perrot, J.-C. (1977). *L'Âge d'or de la statistique régionale française. An IV-1804.* Paris: S.E.R.

Perrot, J.-C. (1992). *Une histoire intellectuelle de l'économie politique. XVIIe-XVIIIe siècle.* Paris: éditions de l'E.H.E.S.S.

Perrot, M., & Robert, P. (eds.). (1989). *Compte général de l'administration de la justice criminelle en France pendant l'année 1880 et rapport relatif aux années 1826 à 1880.* Genève: Slatkine reprints.

Petit, A. (ed.). (2003). *Auguste Comte, trajectoires positivistes, 1798–1998.* Paris: L'Harmattan.

Poggi, S. (1994). Neurology and biology in the romantic age in Germany: Carus, Burdach, Gall, von Baer. In S. Poggi & M. Bossi (eds.), *Romanticism in Science: Science in Europe, 1790–1840.* Dordrecht & Boston: Kluwer Academic Publishers, pp.143–160.

Poisson, D. (1824, 1829). Sur la probabilité des résultats moyens des observations. *Additions pour la Connaissance des tems de l'an 1827*, Paris: Bachelier, pp. 273–302, Suite du Mémoire sur la probabilité du résultat moyen des observations. *Additions pour la Connaissance des tems de l'an 1832*, Paris: Bachelier, pp. 3–22.

Poisson, D. (1830). Mémoire sur la proportion des naissances des filles et des garçons, *Mémoires de l'Académie royale des sciences de l'Institut de France*, Paris: Firmin-Didot, *IX*, pp. 239–308.

Poisson, D. (1835). Recherches sur la probabilité des jugements principalement en matière criminelle. *Comptes-rendus hebdomadaires des séances de l'Académie des sciences...*, Paris: Bachelier, *1*, pp. 473–494.

Poisson, D. (1837). *Recherches sur la probabilité des jugements en matière criminelle et en matière civile. Précédées des Règles générales du calcul des probabilités.* Paris: Bachelier.

Poisson, D., & Nicollet, J. (1824). Observations relatives au nombre de naissances des deux sexes. *Annuaire pour l'an 1825... par le Bureau des longitudes*, Paris: Firmin-Didot, pp. 98–99 (continued as: *id., Annuaire 1826, ibid.*, 1825, pp. 102–103, *id., Annuaire 1827, ibid.*, 1826, pp. 92–93; *etc., id., Annuaire 1832, ibid.*, 1832, 2nd ed., pp. 98–99).

Porter, T. M. (1986). *The rise of statistical thinking, 1820–1900.* Princeton: Princeton University Press.

Porter, T. M. (1995). *Trust in numbers. The pursuit of objectivity in science and public life.* Princeton: Princeton University Press.

Porter, T. M. (2004). *Karl Pearson. The scientific life in a statistical age.* Princeton: Princeton University Press.

Pressat, R. (1979). *Dictionnaire de démographie.* Paris: P.U.F.

Prévost, J.-G. (2002). Genèse particulière d'une science des nombres. L'autonomisation de la statistique en Italie entre 1900 et 1914. *Actes de la recherche en sciences sociales, 141–142,* pp. 98–109.

Qian, N. (2005). Missing women and the price of tea in China: The effect of relative female income on sex imbalance. *Job Market Paper*, MIT, 58.

Quetelet, A. (1828). *Instructions populaires sur le calcul des probabilités.* Bruxelles: Tarlier et Hayez.

Quetelet, A. (1832). *Sur la Possibilité de mesurer l'influence des causes qui modifient les éléments sociaux. Lettre à M. Villermé, de l'Institut de France, etc.* Bruxelles: Hayez.

Quetelet, A. (1835). *Sur l'Homme et le développement de ses facultés ou Essai de Physique sociale.* Paris: Bachelier, *Physique sociale ou Essai sur le développement des facultés de l'Homme.* Bruxelles, Mucquardt - Paris, Baillière - Saint Pétersbourg, Issakoff, 1869 (critical ed.: Bruxelles, Académie royale de Belgique, 1997).

Quetelet, A. (1838). *De l'influence des saisons sur la mortalité aux différents âges dans la Belgique.* Bruxelles: Hayez.

Quetelet, A. (1842). *A treatise on man and the development of his faculities.* Edinburgh: Chambers, 1842.

Quetelet, A. (1846). *Lettres [...] sur la théorie des probabilités appliquée aux sciences morales et politiques.* Bruxelles: Hayez.

Quetelet, A. (ed.). (1853). *Compte rendu des travaux du Congrès général de statistique réuni à Bruxelles les 19, 20, 21 et 22 septembre 1853.* Bruxelles: Hayez.

Quetelet, A. (1873). *Congrès international de statistique. Sessions de Bruxelles (1853), Paris (1855), Vienne (1857), Londres (1860), Berlin (1863), Florence (1867), La Haye (1869) et St-Pétersbourg (1872).* Bruxelles: Hayez.

Quetelet, A., & Heuschling, X. (eds.). (1865). *Statistique internationale (population)....* Bruxelles: Hayez.

Registrar-General [of Great Britain]. (1869). *Twenty-ninth annual report of the Registrar-General for 1866.* London: Registrar-General Office.

Regnault, J. (1936). *Fille ou garçon? Comment avoir fille ou garçon. Diagnostic précoce de la grossesse. Comment connaître le sexe avant la naissance.* Paris: Médicis.

Revel, J. (ed.). (1996). *Jeux d'échelles. La microanalyse à l'expérience.* Paris: Seuil & Gallimard.

Ripley, W. Z. (1908). Sex-ratios at birth in town and country. *The Quarterly Journal of Economics, 23*(1), pp. 175–178.

Roche, D. (1989). *La Culture des apparences. Une histoire du vêtement XVIIe–XVIIIe siècle.* Paris: Fayard.

Roche, D. (1993). *La France des Lumières.* Paris: Fayard.

Rohlfs, C., Reed, A., & Yamada, H. (2005). *Missing women and the year of the fire horse: Changes in the value of girls and child avoidance mechanisms in Japan, 1846, 1906 and 1966*, Working paper, University of Chicago (home.uchicago.edu/~car/firehorse.pdf; June 6th).

Rohrbasser, J.-M. (2001). *Dieu, l'ordre et le nombre. Théologie physique et dénombrement au XVIIIe siècle.* Paris: P.U.F.

Rohrbasser, J. -M. & Brian, É. (2005). Histoire de la preuve par le nombre des naissances. In M. Halbwachs (2005), pp. 73-84.

Rosental, P.-A. (1997). Thirteen years of debate: from population history to French historical demography (1945–1958). *Population: An English Selection*, pp. 215–241 (first publication in French, *Population, 51*(6), 1996, pp. 1211–1238).

Sadler, M. T. (1830). *The law of population. A Treatise, in six books, in disproof of the superfecundity of human beings, and developing the real principle of their increase.* London: Murray, 2 vols. (reprint: London, Routledge & Thoemmes Press, 1994).

Schor, P. (2001). *Compter et classer: histoire des catégories de la population dans le recensement américain 1790–1940.* Paris: Ph. D., E.H.E.S.S.

Schuster, H. (1908). *Johann Gottfried Hoffmann als Nationalökonom. Ein Beitrag zur Geschichte der Nationalökonomie des 19. Jahrhunderts.* Berlin: Walther.

Schwartz, P. (1934). Maurice Halbwachs, Statistische Unter-suchungen über die Geschlechtsbestimmung der Geborenen. *Allgemeines Statistisches Archiv, 23*(3), p. 439.

Schweber, L. (1997). Controverses et styles de raisonnement. Débats sur la statistique de population au XIXe siècle en France et en Angleterre. *Enquête*, (5), pp. 83–108.

Séguy, I. (1997). *Les Aspects religieux et profanes dans le traitement funéraire réservé au nouveau-né au Moyen Âge et à l'époque Moderne.* In L. Buchet (ed.), *L'enfant, son corps et son histoire*, Sophia Antipolis, APDCA, pp. 97–113.

Séguy, I. (et al.). (2001). *La Population de la France de 1670 à 1829. L'Enquête Louis Henry et ses données.* Paris : éditions de l'I.N.E.D.

Simiand, F. (1922). *Statistique et expérience. Remarques de méthode.* Paris : Rivière.

Simiand, F. (1932). *Le Salaire, l'évolution sociale et la monnaie* (3 vols.). Paris: Alcan.

Simmel, G. (1898). Comment les formes sociales se maintiennent. *L'Année sociologique 1896–1897, 1*, pp. 71–109.

Simmel, G. (1902). The number of members as determining the sociological form of the group. *The American Journal of Sociology, 1*(46), pp. 158–196 (later publ. in *Soziologie. Untersuchungen über die Formen der Vergesellschaftung*, Berlin, Duncker et Humblot, 1908; Francfort, Suhrkamp, 1982).

Simon, P., & Clément, M. (2006). How should the diverse origins of people living in France be described? *Population & Societies*, (425).

Sober, E. (1984). *The nature of selection.* Cambridge: M.I.T.-Press.

Sober, E. (2003). Sex Ratio Theory, Ancient and Modern. An 18th Century Debate about Intelligent Design and the Development of Models in Evolutionary Biology, paper for the *Workshop on the History and Philosophy of Artificial Life* organized by Jessica Riskin at Stanford University, (posted in 2005 at philosophy.wisc.edu/sober/papers.htm).

Spallanzani, L. (1785). *Expériences pour servir à l'histoire de la génération des animaux et des plantes.* Genève: Chirol.

Steinmetz, J. (2003). *Wiederentdecken – aufdecken. Johann Peter Süßmilchs Versuch eines Beweises, daß die erste Sprache ihren Ursprung nicht vom Menschen, sondern allein vom Schöpfer erhalten habe (1766) als (sprach)wissenschaftliche Arbeit der Aufklärungszeit.* Hannover: Wehrhahn.

Stieda, W. (1875). *Sexualverhältniss der Geborenen. Eine statistische Studie.* Strassburg (*Statistischen Mitteilungen über Elsass-Lothringen*, Vol. 5).

Stigler, S. M. (1986). *The history of statistics. The measurement of uncertainty before 1900.* Cambridge: Harvard University Press.

Stigler, S. M. (1999). *Statistics on the table. The history of statistical concepts and methods.* Cambridge: Harvard University Press.

Stolkowski, J., & Choukroun, J. (1991). *Garçon ou fille ? Le choix par sélection préconceptionnelle.* Paris: Chiron.

Stolkowski, J. (1984). *Le Secret du paradis: fille ou garçon ? Choisir le sexe de son enfant grâce à une alimentation appropriée.* Montréal: éditions de l'Homme.

Strüssmann, C. A., & Patiño, R. (1999). Sex Determination, Environmental. In Knobil & Neill, Vol. 4, pp. 402–409.

Styrna, J. (1999). Sex Determination, Genetic. In Knobil & Neill, Vol. 4, pp. 409–414.

Süssmilch, J. P. (1741). *Die göttliche Ordnung in den Veränderungen des menschlischen Geschlechts, aus der Geburt, Tod, und Fortpflanzung desselben erwiesen*. Berlin: J. C. Spener (French trans. by J.-M. Rohrbasser, *L'Ordre divin dans les changements de l'espèce humaine, démontré par la naissance, la mort et la propagation de celle-ci*, Paris, éditions de l'I.N.E.D., 1998).

Szreter, S. (1996). *Fertility, class and gender in Britain 1860–1940*. Cambridge: Cambridge University Press.

Tardieu, A. (1868). *Étude médico-légale sur l'infanticide*. Paris: Baillière.

Thury, M. (1863). *Mémoire sur la loi de production des sexes chez les plantes, les animaux et l'homme*. Genève: Librairie Joël Cherbuliez.

Tillier, A. (2001). *Des criminelles au village. Femmes infanticides en Bretagne (1825–1865)*. Rennes: Presses universitaires de Rennes.

Todhunter, I. (1865). *A history of the mathematical theory of probability. From the time of Pascal to that of Laplace*. Cambridge: MacMillan.

Tooze, A. (2001). *Statistics and the German State, 1900–1945. The making of modern economic knowledge*. Cambridge: Cambridge University Press.

Tort, P. (ed.). (1996a). *Dictionnaire du darwinisme et de l'évolution* (3 vols.). Paris: P.U.F.

Tort, P. (1996b). Sélection naturelle (Darwin et la —). In Tort (1996a) pp. 3931–3940.

Tort, P. (1996c). Sélection sexuelle, In Tort (1996a) pp. 3947–3949.

Tourdes, G., & Metzquer, E. (1896). *Traité de médecine légale théorique et pratique*. Paris: Asselin & Houzeau.

Tourdes, G. (1889). Infanticide. In A. Dechambre (ed.), *Dictionnaire encyclopédique des sciences médicales*, Paris: Masson & Asselin, Vol. 15, pp. 612–672.

Trivers, R. L., & Willard, D. E. (1973). Natural selection of parental ability to vary the sex ratio of offspring. *Science, 68*(179), pp. 90–92.

Usui, T., Houhami, J., & Kaneko, I. (1976). Hinoe-Uma Umareno Shihyo [Women's statistics for the year of the Fire Horse] (in Japanese). *Kousei no Shihyo* [*Public Health Statistics*], *23*(3), pp. 3–13.

Vallin, J. (2002). Mortalité, sexe et genre. In Caselli, Vallin & Wunsch, pp. 319–350.

Van de Walle, E. (1974). *The female population of France in the nineteenth century. A reconstruction of 82 départements*. Princeton: Princeton University Press.

Venn, J. (1888). *The logic of chance. An essay on the foundations and province of the theory of probability, with especial reference to its application to moral and social science*. London: Cambridge, Macmillan (3rd ed.).

Veyne, P.(1996). L'interprétation et l'interprète. A propos des choses de religion. *Enquête. Anthropologie, histoire, sociologie*, (3), pp. 241–272.

Villermé, L.-R. (1832a). Compte-rendu de 'De l'effet de la légitimité sur le rapport des naissances de différents sexes', de Pierre Prévost (*Bibliothèque universelle de Genève*, October 1829). *Annales d'hygiène publique et de médecine légale*, Paris, Vol. 7, p. 453.

Villermé, L.-R. (1832b). Compte-rendu de 'Lettre de M. le Professeur Charles Babbage à l'Honorable M. T.-P. Courtenay, sur le rapport des deux sexes dans les naissances' (*The Edinburgh Journal of Science conducted by David Brewster*, 1829), *Annales d'hygiène publique et de médecine légale*, Paris, Vol. 7, pp. 445–453.

Villermé, L.-R. (1832c). Compte-rendu de 'Recherches sur le rapport des deux sexes dans les naissances de M. Bickes' (*Zeitung für das Gesamte Medizinalwesen*, 7 February 1831). *Annales d'hygiène publique et de médecine légale*, Paris, Vol. 7, pp. 453–459.

Vismann, C. (2001). *Akten. Medientechnik und Recht.* Frankfurt: Fischer Taschenbuch Verlag.

Weatherby, J. (1791). *An Introduction to a General Stud-Book* [...]. London.

Weiner, D. (1993). *The Citizen-patient in revolutionary and imperial Paris.* Baltimore: The Johns Hopkins University Press.

Westergaard, H. (1932). *Contributions to the history of statistics.* London: King.

Wietog, J. (2001). *Volkszählungen unter dem Nationalsozialismus. Eine Dokumentation zur Bevölkerungsstatistik im Dritten Reich.* Berlin: Duncker & Humblot.

Williams, G. C. (1975). *Sex and evolution.* Princeton: Princeton University Press.

Wise, N. (ed.). (1994). *The Values of precision.* Princeton: Princeton University Press.

World Health Organization (1978). Sex differentials. In *Social and biological effects on perinatal mortality. Report on an international comparative study sponsored by the World Health Organization* (pp. 77–85). Budapest: Statistical Publishing House.

World Health Organization (2006). Determinants or causes of death. In *Neonatal and Perinatal mortality. Country, regional and global estimates.* Geneva: World Health Organization, pp. 4–5.

Yacine, T. (1988). *L'Izli ou l'amour chanté en kabylie.* Paris: éditions de la Maison des sciences de l'Homme.

Yacine, T. (2006). *Si tu m'aimes, guéris-moi. Etudes d'ethnologie des affects en Kabylie.* Paris: éditions de la Maison des sciences de l'Homme.

Zhang, G. (2004). *China's far-below replacement level fertility. A reality or illusion arising from underreporting of births?*, Canberra: Ph.D., Australian National University.

Zuberi, T. (2001). *Thicker than blood. An essay on how racial statistics lie.* Minneapolis: University of Minnesota Press.

INDEX

METHODOS SERIES

1. R. Franck (ed.): *The Explanatory Power of Models*. 2002
ISBN 978-1-4020-0867-2
2. D. Courgeau (ed.): *Methologoy and Epistemology of Multilevel Analysis*. 2003
ISBN 978-1-4020-1475-8
3. D. Pumain (ed.): *Hierarchy in Natural and Social Sciences*. 2006
ISBN 978-1-4020-4126-6
4. E. Brian and M. Jaisson: *The Descent of Human Sex Ratio at Birth*: A Dialogue between Mathematics, Biology and Sociology. 2007
ISBN 978-1-4020-6035-9

MIX
Papier aus verantwortungsvollen Quellen
Paper from responsible sources
FSC® C105338

If you have any concerns about our products,
you can contact us on
ProductSafety@springernature.com

In case Publisher is established outside the EU,
the EU authorized representative is:
Springer Nature Customer Service Center GmbH
Europaplatz 3, 69115 Heidelberg, Germany

Printed by Libri Plureos GmbH
in Hamburg, Germany